Purpose of Math Shortcuts to Ace the SAT 1* (New SAT*) and the New PSAT/NMSQT

This book contains the most extensive and unique collection of math shortcuts and strategies for the College Entrance Exam (SAT* and PSAT) available.
Other SAT* books offer a review of arithmetic, algebra, geometry and some <u>general</u> strategies. Our book illustrates 34 <u>specific</u> math shortcuts and strategies that <u>supplement</u> other test preparation materials.

Math Shortcuts

to Ace

the SAT

ACHIEVEMENT PUBLISHING, INC.

ISBN 1-882228-00-6

To order additional copies:
call 1-800-247-6553 (BookMasters, Inc.)
Internet: www.bookmasters.com/marktplc/books/00143.htm

For Questions or Comments:
Achievement Publishing, Inc.
P.O. Box 30146
Enterprise, AL 36331
ph. 1-800-499-6284
fax 1-888-581-2211
e-mail: achpub@prodigy.com

Math Shortcuts

to Ace

the SAT

Table of Contents

Divisibility: Shortcuts #1-2

Comparing Fractions: Shortcuts #3-5

Averages: Shortcuts #6-9

Quantitative Comparisons: Shortcuts #10-14

Algebra: Shortcuts #15-17

Geometry: Shortcuts #18-24

Word Problems: Shortcuts #25-34

Divisibility: Shortcuts 1 - 2

Shortcut #1 When a number is divisible by 3

If the sum of the digits of a number is divisible by 3, then that number is divisible by 3.

Note: The word **divisible** indicates that no remainder exists in the answer to a division problem.

Example 1 **Is 1,545 divisible by 3 ?**

Step 1 Add the digits of the number 1,545 .

$1 + 5 + 4 + 5 = 15$

Step 2 If the sum of the digits of any number is divisible by 3, then that number is divisible by 3. The sum of the digits of 1,545 is 15 and 15 is divisible by 3. Therefore, 1,545 is divisible by 3.

$15 \div 3 = 5$, **therefore 1,545 is divisible by 3**

Example 2 **Is 5,602 divisible by 3 ?**

Step 1 Add the digits of 5,602 .

$5 + 6 + 0 + 2 = 13$

Step 2 If the sum of the digits of a number is **not** divisible by 3, then that number is **not** divisible by 3. The sum of the digits of 5,602 is 13 and 13 is **not** divisible by 3. Therefore, 5,602 is **not** divisible by 3.

$3\overline{)13}$ = 4 r1 , **therefore 5,602 is <u>not</u> divisible by 3**

Exercises: **Determine if the number is divisible by 3. Do so without dividing the number by 3.**

1) 7,032 **2)** 15,363 **3)** 3,510 **4)** 70,153 **5)** 6,231 **6)** 3,031

Shortcut #2 Divisibility of numbers with a repeating digit

Two numbers with repeating digits are divisible if both the following two conditions exist: 1) The repeating digit of the dividend is divisible by the repeating digit of the divisor and 2) The number of digits of the dividend is divisible by the number of digits of the divisor.

Note: The **divisor** is the number that is divided into the **dividend**. In the example, $15 \div 3$, 15 is the dividend and 3 is the divisor. The answer to a division problem is the quotient.

Example **444,444,444 is divisible by which number?**

a) 44 b) 22 c) 2,222 d) 222 e) 555

Step 1 Exclude 555 because the digit of 4 in 444,444,444 is **not** divisible by the digit of 5 in 555.

Step 2 Choices a, b, c and d all meet the first condition, so now we will test the second condition. The number 444,444,444 has 9 digits, therefore choose an answer whose number of digits will divide into 9. The choices of 44 and 22 each have two digits. 9 is **not** divisible by 2, so 44 and 22 are **not** correct. The choice 2,222 has four digits and 9 is **not** divisible by 4, so 2,222 is **not** the correct answer.

Step 3 The choice 222 has three digits. 9 is divisible by 3, therefore 444,444,444 is divisible by 222. Thus, the answer is **choice d**.

Exercises:

1) 5,555 is divisible by which number?
a) 111 b) 555 c) 22 d) 55

2) 333,333 is divisible by which number?
a) 1,111 b) 111 c) 3,333 d) 33,333

3) 888,888,888 is divisible by which number?
a) 333 b) 44 c) 888 d) 2,222

4) 444,444 is divisible by which number?
a) 2,222 b) 22,222 c) 22 d) 333

5) 666,666,666 is divisible by which number?
a) 3,333 b) 333 c) 6,666 d) 555

6) 999,999 is divisible by which number?
a) 333 b) 9,999 c) 44 d) 3,333

Comparing Fractions: Shortcuts 3 - 5

Note: When comparing fractions, it is often not necessary and too time consuming to find common denominators.

Shortcut #3 Cross-Multiply to compare two fractions

Multiply the denominator of each fraction by the numerator of the other fraction in an upward direction. The fraction with the greater answer above itself is the greater fraction.

Example **Which fraction is greater?** $\frac{5}{8}$ *or* $\frac{4}{9}$

Step 1 Multiply the denominator of one fraction by the numerator of the other fraction. This multiplication **must** be performed in an upwards diagonal direction. Place each answer above the numerator that was multiplied.

$$\overset{\mathbf{45}}{\frac{5}{8}} \quad \times \quad \overset{\mathbf{32}}{\frac{4}{9}}$$

Step 2 The fraction with the greater number above itself is the greater fraction. **5/8 is greater than 4/9** because 45 is greater than 32.

$$\overset{\mathbf{45}}{\frac{5}{8}} \quad > \quad \overset{\mathbf{32}}{\frac{4}{9}}$$

Exercises: **Which fraction is greater? (Hint: Use cross-multiplication)**

1) $\frac{2}{3}$ *or* $\frac{3}{4}$ **2)** $\frac{7}{10}$ *or* $\frac{5}{9}$ **3)** $\frac{4}{7}$ *or* $\frac{3}{5}$ **4)** $\frac{3}{7}$ *or* $\frac{4}{9}$

Shortcut #4 If each numerator is 1 less than its denominator

If each numerator is 1 less than its denominator, then the fraction with the greatest numerator is the greatest fraction.

Example **Which fraction is greatest?** $\frac{1}{2} \quad \frac{2}{3} \quad \frac{4}{5} \quad \frac{5}{6}$

Step 1 Notice that each numerator is 1 less than the denominator. If this pattern exists, then the greatest fraction is the one with the greatest numerator. **5/6** has the greatest numerator, therefore it is the greatest fraction.

Exercises: **Which fraction is greatest?**

1) $\frac{1}{2} \quad \frac{3}{4} \quad \frac{4}{5} \quad \frac{7}{8}$ **2)** $\frac{3}{4} \quad \frac{5}{6} \quad \frac{10}{11} \quad \frac{12}{13}$ **3)** $\frac{11}{12} \quad \frac{1}{2} \quad \frac{9}{10} \quad \frac{2}{3}$

Shortcut #5 Compare each fraction to ½

Example **Which fraction is greatest?** $\frac{3}{8} \quad \frac{4}{7} \quad \frac{4}{9} \quad \frac{5}{11}$

Step 1 Compare each fraction to ½. This can be done quickly by determining if the numerator is more or less than half the denominator. The only fraction whose numerator is more than half the denominator is 4/7. Therefore, **4/7** is the greatest fraction.

Exercises: **Which fraction is greatest?**

1) $\frac{1}{3} \quad \frac{5}{9} \quad \frac{4}{10} \quad \frac{6}{13}$ **2)** $\frac{1}{2} \quad \frac{2}{5} \quad \frac{2}{3} \quad \frac{3}{8}$ **3)** $\frac{3}{7} \quad \frac{1}{3} \quad \frac{3}{6} \quad \frac{5}{8}$

Averages: Shortcuts 6 - 9

Shortcut #6 When the average is the middle number

The average of a group of numbers that increase by the same amount is the middle number.

Example 1 **Find the average of 7, 8, 9, 10, 11.**

Step 1 The numbers increase by the same amount, 1. The average is therefore the middle number, **9**.

Example 2 **Find the average of 25, 30, 35.**

Step 1 The numbers increase by the same amount, 5. The average is therefore the middle number, **30**.

Example 3 **Find the average of 20, 30, 40, 50.**

Step 1 The numbers increase by the same amount, 10. The average is therefore the middle number, **35**.

Note: The middle number (and average) does **not** have to be listed in the question.

Exercises: **Find the average.**

1) 6, 7, 8, 9, 10 **2)** 10, 11, 12, 13, 14 **3)** 6, 7, 8, 9

4) 30, 40, 50 **5)** 30, 40, 50, 60 **6)** 15, 18, 21, 24, 27

7) 3, 6, 9, 12 **8)** 20, 40, 60, 80 **9)** 20, 40, 60, 80, 100

10) 5, 10, 15, 20, 25, 30 **11)** 8, 12, 16, 20, 24, 28

Shortcut #7 When the average equals the average of the first and last number

The average of a group of numbers that increase by the same amount is the average of the first and last numbers.

Note: **Shortcuts 7 and 8 apply to the same type of problem. It is quicker to use shortcut 7 when determining the average of a few numbers and use shortcut 8 when determining the average of many numbers.**

Example **Find the average of the numbers 1 through 70 inclusive.**

Note: The word **"inclusive"** means that the first and last numbers are **included,** thus the numbers 1 through 70 inclusive, includes 1 and 70.

Step 1 Because these numbers increase by the same amount, 1, the average of all these numbers will equal the average of the first and last number.

$$\frac{1 + 70}{2} = \frac{71}{2} = \mathbf{35\frac{1}{2}}$$

Exercises: **Find the average.**

1) 1 through 50 inclusive **2)** 1 through 100 inclusive **3)** 1 through 1000 inclusive

4) 1 through 2000 inclusive **5)** 1 through 76 inclusive **6)** 1 through 500 inclusive

Shortcut #8 Determining the sum of consecutive numbers

The sum of consecutive numbers is found by multiplying their average by the number of consecutive numbers.

Note: **Consecutive numbers** are numbers ordered one after the other without interruption. Examples of consecutive numbers are: (1, 2, 3, 4) or (7, 8, 9) or (7, 8, 9...). Notice that consecutive numbers may or may not continue indefinitely.

Example **Calculate the sum of the numbers 1 through 70 inclusive.**

Step 1 The numbers 1 through 70 inclusive (1, 2...69, 70), are consecutive numbers. Therefore their sum will equal their average multiplied by how many numbers exist in the group, 70. The average of the numbers 1 through 70 inclusive was found to be 35 ½ in shortcut #8.

Step 2 Multiply the average (35 ½) by the amount of numbers, 70. This answer will equal the sum of the numbers 1 through 70 inclusive.

$$\left(35\frac{1}{2}\right)(70) =$$

$$\left(\frac{71}{2}\right)(70) =$$

cancel: $$\frac{(71)}{(\not{2})_{1}}(\not{70}^{35}) = \mathbf{2,485}$$

Note: Multiplication is indicated if a math operation does **not** exist. For example, (3) (5) represents 3 x 5 .

Exercises: **Calculate the sum in the exercises from shortcut # 8**

Shortcut #9 When each value could equal the average

When determining the sum of a group of numbers, assume each number equals the average.

Example **The maximum allowable weight in an elevator is 1,500 pounds. Three people with an average weight of 100 pounds are in the elevator. How many more pounds is the elevator allowed to carry?**

Step 1 Determine the sum of the weight of the three people. It can be assumed that each person's weight equals the average of 100 pounds, because their sum is to be determined. Thus, the elevator now carries 300 pounds.

$$100 + 100 + 100 = 300 \text{ pounds}$$

Step 2 Subtract 300 from 1,500 to determine the additional weight allowable in the elevator.

$$1{,}500 - 300 = \mathbf{1{,}200 \text{ pounds}}$$

Exercises:

1) The maximum allowable weight in an elevator is 2,000 pounds. Four people with an average weight of 140 pounds are in the elevator. How many more pounds is the elevator allowed to carry?

2) The maximum allowable weight in an elevator is 3,000 pounds. Five people with an average weight of 160 pounds are in the elevator. How many more pounds is the elevator allowed to carry?

3) The maximum allowable weight a truck may hold is 5,000 pounds. Ten boxes with an average weight of 150 pounds are already on the truck. How many more pounds is the truck allowed to carry?

Quantitative Comparisons: Shortcuts 10 - 14

The possible answers to comparison questions are: A, B, C, or D. For choice A, B or C to be the answer, it must <u>always</u> be true. If choice A, B or C is <u>sometimes</u> but not <u>always</u> true, the answer is choice D.
The directions for comparison questions are as follows:

SUMMARY DIRECTIONS FOR COMPARISON QUESTIONS

<u>Answer:</u> **A** if the quantity in Column A is greater;
B if the quantity in Column B is greater;
C if the two quantities are equal;
D if the relationship cannot be determined from the information given.

AN E RESPONSE WILL NOT BE SCORED.

Note: Information may be given in individual questions describing possible values for variables. In a given question, a symbol that appears in both columns represents the same thing in Column A as it does in Column B. Letters such as x, n, and k stand for real numbers.

Shortcut #10 Compare pieces

If variables (letters) do <u>not</u> exist in the comparison, then the answer may often be determined by comparing the individual pieces (terms) that are separated by an addition or subtraction sign. The same math operation (+ or –) <u>must</u> precede the individual pieces (terms) that are compared.

Hint: The answer to comparisons that do <u>not</u> contain variables (letters) is either A, B or C. Choice D is <u>not</u> a possible answer if the comparison does not contain a variable (letter).

SUMMARY DIRECTIONS FOR COMPARISON QUESTIONS

<u>Answer:</u> **A** if the quantity in Column A is greater;
B if the quantity in Column B is greater;
C if the two quantities are equal;
D if the relationship cannot be determined from the given information.

AN "E" RESPONSE WILL NOT BE SCORED.

Example 1

Column A	Column B
$\sqrt{7} + \sqrt{6}$	$\sqrt{5} + \sqrt{3}$

Step 1 Both column A and column B do **not** contain variables, therefore a comparison by pieces should be attempted. Notice that all four terms are preceded by the same math operation (addition signs in this case), thus any term from one column may be compared from any term from the other column.

Step 2 $\sqrt{7}$ is greater than $\sqrt{5}$ and $\sqrt{6}$ is greater than $\sqrt{3}$. **Column A** is greater because each piece in Column A is greater than a particular piece in Column B.

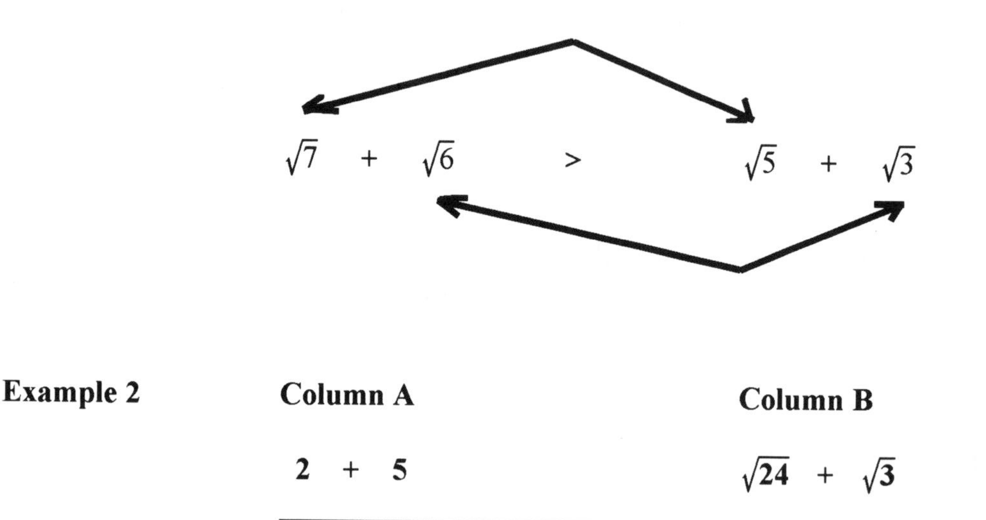

Example 2

Column A	Column B
$2 + 5$	$\sqrt{24} + \sqrt{3}$

Step 1 Do **not** add $2 + 5$ until a comparison by pieces is attempted. A comparison by pieces is possible because both column A and column B do **not** contain variables. Notice that all four terms are preceded by the same math operation (addition signs in this case), thus any term from one column may be compared with any term from the other column.

Step 2 Compare 2 with $\sqrt{3}$ and compare 5 with $\sqrt{24}$. 2 is greater than $\sqrt{3}$ because 2 equals $\sqrt{4}$. 5 is greater than $\sqrt{24}$ because 5 equals $\sqrt{25}$. **Column A** is therefore greater than Column B because each piece in Column A is greater than a particular piece in Column B.

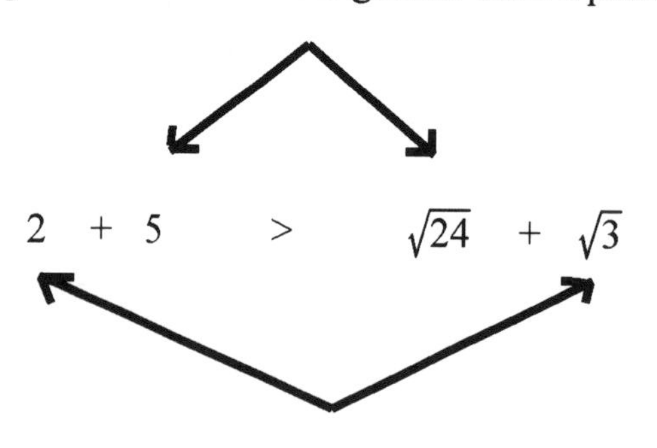

Exercises: SUMMARY DIRECTIONS FOR COMPARISON QUESTIONS

Answer:
- **A** if the quantity in Column A is greater;
- **B** if the quantity in Column B is greater;
- **C** if the two quantities are equal;
- **D** if the relationship cannot be determined from the information given.

AN "E" RESPONSE WILL NOT BE SCORED.

	Column A	**Column B**
1)	$\sqrt{2} + \sqrt{3}$	$\sqrt{5} + \sqrt{6}$
2)	$\sqrt{7} + \sqrt{2}$	$\sqrt{5} + \sqrt{10}$
3)	$\sqrt{12} + \sqrt{7}$	$\sqrt{10} + \sqrt{6}$
4)	$\sqrt{3} + \sqrt{7}$	$\sqrt{5} + \sqrt{2}$
5)	$\sqrt{6} + \sqrt{12}$	$\sqrt{13} + \sqrt{7}$
6)	$\sqrt{63} + \sqrt{8}$	$8 + 3$
7)	$\sqrt{5} + \sqrt{18}$	$2 + 4$
8)	$3 + 5$	$\sqrt{8} + \sqrt{24}$
9)	$7 + 2$	$\sqrt{50} + \sqrt{5}$
10)	$\sqrt{4} + \sqrt{9}$	$2 + 3$

Shortcut #11 Eliminate identical pieces

Eliminate identical pieces by performing the same math operation on the quantity in each column. Never multiply nor divide by a negative number, zero or a variable that could be negative or zero.

Note: If variables remain, then substitute numbers as shown in Shortcut #13.

SUMMARY DIRECTIONS FOR COMPARISON QUESTIONS

<u>Answer:</u>
- **A** if the quantity in Column A is greater;
- **B** if the quantity in Column B is greater;
- **C** if the two quantities are equal;
- **D** if the relationship cannot be determined from the information given.

AN "E" RESPONSE WILL NOT BE SCORED.

Example 1

Column A	**Column B**
n + 1	**n + 5**

Step 1 Subtract "n" from each column.

$n + 1 - n$		$n + 5 - n$
1	<	5

Step 2 Compare the column values. 5 is greater than 1, therefore the answer is **choice B**.

Example 2

Column A	**Column B**
5.6 x 732	**5 x 732**

Hint: The answer to comparisons that do <u>not</u> contain variables (letters) will be either A, B or C.

Step 1 Divide the quantity in each column by 732. Then cancel 732 in each column.

$\frac{5.6 \times 732}{732}$		$\frac{5 \times 732}{732}$
$\frac{5.6 \times \cancel{732}}{\cancel{732}}$		$\frac{5 \times \cancel{732}}{\cancel{732}}$
5.6	>	5

Step 2 Compare the column values. 5.6 is greater than 5, therefore the answer is **choice A**.

Example 3

	Column A	Column B
	$\sqrt{2} + \sqrt{10}$	$\sqrt{7} + \sqrt{2}$

Step 1 Subtract $\sqrt{2}$ from each column.

$$\sqrt{2} + \sqrt{10} - \sqrt{2} \qquad \sqrt{7} + \sqrt{2} - \sqrt{2}$$

$$\sqrt{10} \quad > \quad \sqrt{7}$$

Step 2 Compare the column values. $\sqrt{10}$ is greater than $\sqrt{7}$, therefore the answer is **choice A**.

Exercises: SUMMARY DIRECTIONS FOR COMPARISON QUESTIONS

Answer:
- **A** if the quantity in Column A is greater;
- **B** if the quantity in Column B is greater;
- **C** if the two quantities are equal;
- **D** if the relationship cannot be determined from the information given.

AN "E" RESPONSE WILL NOT BE SCORED.

	Column A	Column B
1)	$n + 2$	$n + 6$
2)	$n - 1$	$n - 2$
3)	$6 - n$	$10 - n$
4)	n	$n + 5$
5)	$\frac{1}{3} + \frac{2}{5}$	$\frac{1}{3} + \frac{4}{5}$
6)	$\frac{6}{7} + \frac{1}{5}$	$\frac{1}{5} + \frac{1}{3}$
7)	4×35	4.1×35
8)	203×2	1.9×203

	Column A	Column B
9)	$\sqrt{7} + \sqrt{2}$	$\sqrt{7} + \sqrt{5}$
10)	$\sqrt{3} + \sqrt{2}$	$\sqrt{5} + \sqrt{3}$
11)	$\sqrt{6} + \sqrt{5}$	$\sqrt{6} + \sqrt{3}$

Shortcut #12 Simplify and then substitute numbers for variables

1st) If possible, simplify the quantity in either or both columns. Some common simplification techniques are: 1) cancellation 2) factor and then cancel 3) distributive multiplication (in some cases) 4) apply an exponent; $(3x)^2 = 9x^2$.

2nd) If possible, simplify the quantities in each column by performing the same math operation on the quantity in each column. Never multiply nor divide both columns by a negative number, zero or a quantity that could be negative or zero.

Note: Multiplication or division of both columns by a variable or variable expression is generally not a great time saver. In addition, the test taker runs the risk of violating the exception. For these reasons, this procedure will be mentioned as an alternative step.

3rd) If a variable still remains, substitute numbers for the variable. The numbers that should be substituted are: 0, 1, −1, a positive integer and a negative integer. If a variable is in a denominator or has an exponent, a fraction should be substituted as well. If a column is not always greater than the other column or if the columns are not always equal, the answer is choice D.

Note: Simplification is **not** necessary prior to number substitution, however it will simplify numerical calculations. In some cases, immediate number substitution may be quicker than simplification.

Note: If any number is substituted for a variable in a denominator and the value of the denominator becomes zero, then the substitution of this number is **not** allowed. This is because the value of a fraction whose denominator is zero is undefined.

SUMMARY DIRECTIONS FOR COMPARISON QUESTIONS

Answer:
- **A** if the quantity in Column A is greater;
- **B** if the quantity in Column B is greater;
- **C** if the two quantities are equal;
- **D** if the relationship cannot be determined from the information given

AN "E" RESPONSE WILL NOT BE SCORED.

Example 1

Column A		Column B
	$x \geq 1$	
x		x^2

Step 1 Substitute 3 for x.

3		3^2
3	$<$	9

Step 2 Now determine whether or not Column B is **always** greater than Column A. Attempt to find a number for x that will create a different result. If Column B is sometimes, but **not** always greater than Column A, not enough information is given and the answer is Choice D. Substitute 1 for x.

1		1^2
1	$=$	1

Because the results in step 1 and step 2 are different, the answer is **Choice D.**

Note: Both columns could initially have been divided by x. This step is valid because x can never be negative nor zero.

Example 2

Column A		Column B
	x is an integer, $x \neq 1$, $x \neq 0$	
x		x^3

Step 1 Substitute 2 for x.

2		2^3
2	$<$	8

Step 2 Attempt to find a number for x that will create a different result.
Substitute -2 for x. Notice that $(-2)^3 = (-2)(-2)(-2) = -8$.

-2		$(-2)^3$
-2	$>$	-8

Because the results in step 1 and step 2 are different, the answer is **Choice D.**

Note: Both columns could **not** have been divided by x, because x could be negative.

Example 3

Column A		**Column B**
	$x \neq 1,\ x > 0$	
x		x^2

Step 1 Substitute 2 for x.

2		$(2)^2$
2	$<$	4

Step 2 Attempt to find a number for x that will create a different result. Substitute ½ for x.

$\frac{1}{2}$		$(\frac{1}{2})^2$

$\frac{1}{2} > \frac{1}{4}$

Because the results in step 1 and step 2 are different, the answer is **Choice D**.

Note: Both columns could initially have been divided by x, because x can **not** be negative nor zero.

Example 4

Column A		**Column B**
	$x > 0,\ x \neq 1$	
$\frac{1}{x}$		x

Step 1 Substitute 2 for x.

$\frac{1}{2}$	$<$	2

Step 2 Attempt to find a number for x that will create a different result. Substitute ½ for x.

Note: The original horizontal line represents the division sign (in this example the original horizontal line is in the fraction $\frac{1}{x}$. If the original horizontal line can not be determined, then the longer horizontal line represents the division sign.

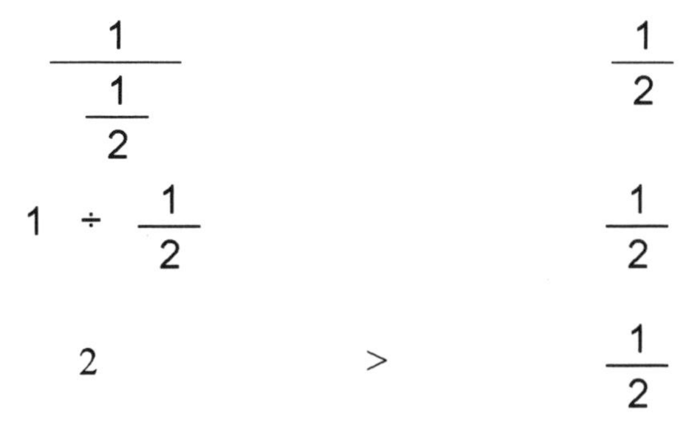

$$\frac{1}{\frac{1}{2}} \qquad \frac{1}{2}$$

$$1 \div \frac{1}{2} \qquad \frac{1}{2}$$

$$2 \quad > \quad \frac{1}{2}$$

Because the results in step 1 and step 2 are different, the answer is **Choice D.**

Example 5

Column A		**Column B**
	$3n - 2 = 6 - n$	
$3n$		$n + 4$

Step 1 Solve the equation for n.

$$3n - 2 + n = 6 - n + n$$

$$4n - 2 = 6$$

$$4n - 2 + 2 = 6 + 2$$

$$4n = 8$$

$$\frac{4n}{4} = \frac{8}{4}$$

$$n = 2$$

Step 2 n always equals 2, therefore 2 is the only number that can be substituted for n.

$3n$		$n + 4$
$3(2)$		$2 + 4$
6	$=$	6

The columns are always equal, therefore the answer is **Choice C.**

Example 6

	Column A		Column B
		$x \neq 0$	
	2x		**x**

Step 1 Simplify by subtracting x from each column.

2x − x		x − x
x		0

Step 2 Substitute 1 for x.

1	>	0

Step 3 Attempt to find a number for x that will create a different result. Substitute −1 for x.

−1	<	0

Because the results in step 1 and step 2 are different, the answer is **Choice D.**

Note: Each column could **not** have been divided by x, because x could be negative.

Example 7

	Column A		Column B
		$x \geq 0$	
	x		**3x**

Step 1 Simplify each quantity by subtracting x.

x − x		3x − x
0		2x

Step 2 Substitute 1 for x.

0		(2) (1)
0	<	2

Step 3 Attempt to find a number for x that will create a different result. Substitute 0 for x.

0		(2) (0)
0	=	0

Because the results in step 1 and step 2 are different, the answer is **Choice D.**

Note: Each column could **not** have been divided by x, because x could be zero.

Example 8

Column A		Column B
	$n \geq 0$	
$2n + 5$		$3n + 1$

Step 1 Subtract 2n from each column.

$2n + 5 - 2n$	$3n + 1 - 2n$
5	$n + 1$

Step 2 Simplify further by subtracting 1 from each column.

$5 - 1$	$n + 1 - 1$
4	n

Step 3 Substitute 1 for n.

$$4 > 1$$

Step 4 Attempt to find a number for n that will create a different result. Substitute 5 for n.

$$4 < 5$$

Because the results in step 1 and step 2 are different, the answer is **Choice D.**

Example 9

Column A		Column B
	$x > 0$	
$5(2x - 3)$		$3(x + 2)$

Step 1 Perform distributive multiplication in each column.

$10x - 15$	$3x + 6$

Step 2 Subtract 3x from each column.

$10x - 15 - 3x$	$3x + 6 - 3x$
$7x - 15$	6

Step 3 Add 15 to each column.

$7x - 15 + 15$	$6 + 15$
$7x$	21

Step 4 Divide each column by 7.

$\frac{7x}{7}$ $\frac{21}{7}$

x 3

Step 5 The given information states that x > 0, thus x could be greater than, less than or equal to 3. Therefore, the answer is **Choice D.**

Example 10

Column A		**Column B**
	x > 0	
$\mathbf{5x^2 + 16}$		$\mathbf{(3x)^2}$

Step 1 Square (3x). Each factor inside the parentheses must be squared, thus $(3x)^2 = 9x^2$.

$5x^2 + 16$ $9x^2$

Step 2 Subtract $5x^2$ from each column.

$5x^2 + 16 - 5x^2$ $9x^2 - 5x^2$

16 $4x^2$

Step 3 Divide both columns by 4. This step is allowed because 4 is positive.

$\frac{16}{4}$ $\frac{4x^2}{4}$

4 x^2

Step 4 Take the square root of both columns. Note that the radical sign ($\sqrt{\ }$) indicates the positive square root only. This step is optional because number substitution at this point would not be too difficult.

$\sqrt{4}$ $\sqrt{x^2}$

2 x

Step 5 Substitute 1 for x.

2 > 1

Step 6 Attempt to find a number for x that will create a different result. Substitute 3 for x.

2 < 3

Because the results in step 5 and step 6 are different, the answer is **Choice D**.

Example 11

Column A	Column B
$\frac{10x - 60}{2}$	$5x - 30$

Step 1 Simplify the quantity in Column A by dividing both terms in the numerator by the denominator.

$$5x - 30 \quad = \quad 5x - 30$$

Both columns are always equal, therefore the answer is **Choice C**.

Example 12

Column A		Column B
	$x > 0$	
$\frac{x^2 + 2x}{x}$		x

Step 1 Simplify the quantity in Column A by dividing each term in the numerator by the denominator.

$$x + 2 \qquad x$$

Step 2 Column A will always be 2 greater than Column B, therefore the answer is **Choice A**. This answer may be made clearer by subtracting x from both columns.

$$x + 2 - x \qquad x - x$$

$$2 \quad > \quad 0$$

It is now clearer that Column A is always greater than Column B, because 2 is always greater than 0.

Simplification in the following two examples requires factoring prior to cancellation because the denominators contain more than one term.
The factoring steps should be applied in the following order:

1st) If a common factor exists, factor out the common factor

2nd) If an expression contains 2 or 3 terms and one of these terms has variable with an exponent of 2, then attempt to factor using double parentheses

Example 13 **Column A** **Column B**

$x \neq 1,\ x \neq -2$

Column A	Column B
$\dfrac{3x - 3}{x - 1}$	$\dfrac{x^2 - x - 6}{x + 2}$

Step 1 In order to cancel, the numerator and denominator must be in factored form. In each column, only the numerator must be factored because each denominator is already in factored form. Recall the factoring steps: 1st) factor out a common factor if possible 2nd) factor using double parentheses if possible.

Column A	Column B
$\dfrac{3(x - 1)}{x - 1}$	$\dfrac{(x - 3)(x + 2)}{x + 2}$

Step 2 Cancel identical factors. In Column A, the factor $(x - 1)$ can be canceled. In Column B, the factor $(x + 2)$ can be canceled.

Column A	Column B
3	$x - 3$

Step 3 To simplify further, add 3 to each column.

Column A	Column B
$3 + 3$	$x - 3 + 3$
6	x

Step 4 x can be any number except 1, therefore x could be greater than, less than or equal to 6. Because there are different results, the answer is **Choice D**.

Example 14 **Column A** **Column B**

$x > 2$

Column A	Column B
$\dfrac{3x^2 + 6x - 24}{x - 2}$	$x + 4$

Step 1 In order to cancel, the numerator and denominator must be in factored form. In Column A, only the numerator must be factored because the denominator is already in factored form. Recall the factoring steps: 1st) factor out a common factor if possible 2nd) factor using double parentheses if possible.

$$\frac{3(x^2 + 2x - 8)}{x - 2} \qquad x + 4$$

$$\frac{3(x + 4)(x - 2)}{x - 2} \qquad x + 4$$

Step 2 Cancel the factor $(x - 2)$ in the numerator and denominator in Column A.

$$3(x + 4) \qquad x + 4$$

Step 3 The quantity in each column may be divided by $(x + 4)$ because $(x + 4)$ can **not** be negative nor zero. Then cancel the factor $(x + 4)$ in each column.

$$\frac{3(x + 4)}{x + 4} \qquad \frac{x + 4}{x + 4}$$

$$3 \qquad > \qquad 1$$

3 is always greater than 1, therefore Column A is always greater than Column B and so the answer is **Choice A**.

Note: If the division by $(x + 4)$ were **not** allowed or if you prefer not to divide each column by $(x + 4)$, the alternative would be to substitute numbers immediately or to perform distributive multiplication and simplify further as shown in Example 8 and 9.

Example 15

Column A	**Column B**
5 (x + 1)	**3 (x + 1)**

Step 1 Each column can **not** be divided by $(x + 1)$ because $(x + 1)$ could be negative or zero. The quantities in each column could be simplified by performing distributive multiplication and simplifying further as shown in Example 8 and 9. However, the answer can be found quicker by immediately substituting numbers. Substitute 0 for x.

$$5(0 + 1) \qquad 3(0 + 1)$$

$$5(1) \qquad 3(1)$$

$$5 \qquad > \qquad 3$$

Step 2 Attempt to find a value for x that will create a different result. Substitute -1 for x.

$$5(-1 + 1) \qquad 3(-1 + 1)$$

$$5(0) \qquad 3(0)$$

$$0 \qquad = \qquad 0$$

Because the results in step 1 and step 2 are different, the answer is **Choice D**.

Exercises: SUMMARY DIRECTIONS FOR COMPARISON QUESTIONS

Answer: **A** if the quantity in Column A is greater;
B if the quantity in Column B is greater;
C if the two quantities are equal;
D if the relationship cannot be determined from the information given.

AN "E" RESPONSE WILL NOT BE SCORED.

	Column A		Column B
1)	n	$n \geq 1$	n^2
2)	n^2	$n \geq 1$	n^3
3)	n^3	n is an integer, $n \neq 1,\ n \neq 0$	n
4)	n^3	n is an integer, $n \neq 1,\ n \neq 0$	n^2
5)	n	$n > 0,\ n \neq 1$	n^2
6)	n	$n > 0,\ n \neq 1$	n^3
7)	$\frac{2}{n}$	$n > 0,\ n \neq 1$	n
8)	$\frac{1}{n}$	$n > 0,\ n \neq 1$	n
9)	$3n$	$2n + 5 = 5n - 1$	$n + 3$
10)	$2n - 1$	$2n = 7n - 15$	5
11)	$3n$	$n \neq 0$	n
12)	$3n$	$n \neq 0$	$5n$

	Column A		Column B
13)	$5n$	$n \geq 0$	$3n$
14)	$2n$	$n \geq 0$	n
15)	$2n + 7$	$n \geq 0$	$3n + 2$
16)	$3n + 10$	$n \geq 0$	$4n + 2$
17)	$3(5n - 2)$	$n > 0$	$2n + 7$
18)	$-2(n - 2)$	$n > 0$	$3(2n - 4)$
19)	$3x^2 + 13$	$x > 0$	$(4x)^2$
20)	$x^2 + 27$	$x > 0$	$(2x)^2$
21)	$\frac{6x - 4}{2}$		$3x - 2$
22)	$5x + 6$		$\frac{15x + 18}{3}$
23)	$\frac{x^2 + 3x}{x}$	$x > 0$	x
24)	$3x$	$x > 0$	$\frac{x^2 - 6x}{x}$
25)	$\frac{2x - 2}{x - 1}$	$x \neq 1$	2
26)	$\frac{5x + 10}{x + 2}$	$x \neq -2,\ x \neq -1$	$\frac{x^2 - 2x - 3}{x + 1}$
27)	$\frac{3x^2 - 3x - 18}{x + 2}$	$x \neq -2$	x
28)	$\frac{2x^2 - 4x - 30}{x - 5}$	$x \neq 5$	$x + 3$

	Column A	Column B
29)	$3(x + 1)$	$2(x + 1)$
30)	$4(2x - 1)$	$5(2x - 1)$

Shortcut #13 Square both quantities in order to eliminate the square roots

If each column contains only 1 term and either or both columns contain a square root, then square (apply an exponent of 2) the entire quantity in each column. Squaring a square root will eliminate the square root symbol, for example: $(\sqrt{5})^2 = 5$.

Note: If either or both columns contain 2 terms, square roots can also be eliminated by squaring the quantity in each column. However, estimation of square roots can be a quicker method in this case as illustrated in Shortcut #15.

Note: A **term** is a group of numbers or variables (letters) that are multiplied together. Terms are separated by an addition or subtraction sign.

For example:	**One term:**	**Two terms:**
	3	$2 + \sqrt{3}$
	$3\sqrt{5}$	$2\sqrt{5} - 3$
	$6n$	$n + 2$

SUMMARY DIRECTIONS FOR COMPARISON QUESTIONS

Answer:
- **A** if the quantity in Column A is greater;
- **B** if the quantity in Column B is greater;
- **C** if the two quantities are equal;
- **D** if the relationship cannot be determined from the information given.

AN "E" RESPONSE WILL NOT BE SCORED.

Example 1	**Column A**	**Column B**
	$3\sqrt{2}$	$\sqrt{17}$

Step 1 Each column contains only 1 term, thus if the quantities in each column are squared, the square root symbols will be eliminated. Notice that the "3" in Column A **must** be squared as well.

$(3\sqrt{2})^2$	$(\sqrt{17})^2$
$9\sqrt{4}$	17
(9) (2)	17
18 >	**17**

Step 2 18 is greater than 17, therefore $3\sqrt{2}$ is greater than $\sqrt{17}$. Thus, the answer is **choice A**.

Example 2

Column A	**Column B**
5	**$2\sqrt{3}$**

Step 1 Each column contains only one term, thus if the quantities in each column are squared, the square root symbols will be eliminated. Notice that the 5 and the 2 **must** be squared as well.

$(5)^2$	$(2\sqrt{3})^2$
25	$(4)(\sqrt{9})$
25	(4) (3)
25 >	12

Step 2 25 is greater than 12, therefore 5 is greater than $2\sqrt{3}$. Thus, the answer is **choice A.**

Example 3

Column A		**Column B**
	x > 0	
$3\sqrt{x}$		**$\sqrt{2x}$**

Step 1 Number substitution could be used immediately, however, squaring both quantities first will be quicker.

$(3\sqrt{x})^2$	$(\sqrt{2x})^2$
9x	2x

Step 2 Subtract 2x from each quantity.

9x - 2x	2x - 2x
7x	0

Step 3 x can only be positive and any positive number substituted for x will make 7x greater than 0. Therefore, Column A will always be greater than Column B so the correct answer is choice A.

Exercises: SUMMARY DIRECTIONS FOR COMPARISON QUESTIONS

Answer:
- **A** if the quantity in column A is greater;
- **B** if the quantity in column B is greater;
- **C** if the two quantities are equal;
- **D** if the relationship cannot be determined from the information given.

AN "E" RESPONSE WILL NOT BE SCORED.

	Column A		Column B
1)	$\sqrt{6}$		$2\sqrt{3}$
2)	$\sqrt{12}$		$2\sqrt{3}$
3)	$5\sqrt{3}$		$3\sqrt{5}$
4)	4		$\sqrt{8}$
5)	2		$\sqrt{5}$
6)	$5\sqrt{2}$		7
7)	10		$7\sqrt{2}$
8)	$2\sqrt{x}$	$x > 0$	$\sqrt{3x}$
9)	$\sqrt{5x}$	$x > 0$	$3\sqrt{x}$
10)	$3\sqrt{2x}$	$x > 0$	$2\sqrt{3x}$

Shortcut #14 Estimating a square root

Estimate a square root by comparing it to the square root of a perfect square.

Note: This shortcut applies if either or both columns contain 2 terms. If each column contains only 1 term, then square each quantity in order to eliminate the square roots as shown in Shortcut #14.

Note: **Perfect squares** are numbers whose square root equals a whole number. The perfect squares are: 1, 4, 9, 16, 25. . . because $\sqrt{1} = 1$, $\sqrt{4} = 2$, $\sqrt{9} = 3$, $\sqrt{16} = 4$, $\sqrt{25} = 5$.

Note: In general, the best estimate will be the square root of a perfect square whose value is nearest the original square root. The estimate may be greater or less than the original square root. Square roots that are in the **same** column must be estimated in the same manner: either both estimates must be higher or both estimates must be lower than the original values.

SUMMARY OF DIRECTIONS OF COMPARISON QUESTIONS

Answer:
A if the quantity in Column A is greater;
B if the quantity in Column B is greater;
C if the two quantities are equal;
D if the relationship cannot be determined from the information given.

AN "E" RESPONSE WILL NOT BE SCORED.

Example	**Column A**	**Column B**
	$\sqrt{10} + \sqrt{5}$	$\sqrt{15}$

Note: Square roots of different numbers can **not** be added or subtracted. However, any math operation inside the **same** square root may be performed.
Therefore, $\sqrt{10} + \sqrt{5} \neq \sqrt{15}$, however, $\sqrt{10 + 5} = \sqrt{15}$.

Step 1 Estimate the square roots as follows: $\sqrt{10}$ as $\sqrt{9}$, $\sqrt{5}$ as $\sqrt{4}$ and $\sqrt{15}$ as $\sqrt{16}$.

Step 2 Perform the comparison using the estimated values. Notice that the original value of column A is greater than the its estimate and the original value of column B is less than its estimate.

(original)	$\sqrt{10} + \sqrt{5}$		$\sqrt{15}$
(estimate)	$\sqrt{9} + \sqrt{4}$		$\sqrt{16}$
	$3 + 2$	$>$	4
	5	$>$	4

Step 3 The original value of column A is greater than 5 and the original value of column B is less than 4. Therefore, column A is greater than column B and so the answer is **choice A**.

Alternate Solution by squaring each quantity:

If either or both columns contain 2 or less terms, square roots can always be eliminated by squaring the quantity in each column. In some cases, each quantity may have to squared twice. However, square root estimation is a quicker method and generally applies on this exam.

Example	**Column A**	**Column B**
	$\sqrt{10} + \sqrt{5}$	$\sqrt{15}$

Step 1 Square each quantity. Notice that $\sqrt{10}$ and $\sqrt{5}$ can **not** be squared individually, but must be squared as a group.

Column A	Column B
$(\sqrt{10} + \sqrt{5})^2$	$(\sqrt{15})^2$
$(\sqrt{10} + \sqrt{5})(\sqrt{10} + \sqrt{5})$	15
$10 + \sqrt{50} + \sqrt{50} + 5$	15
$10 + 2\sqrt{50} + 5$	15
$15 + 2\sqrt{50}$	15

Step 2 Simplify by subtracting 15 from the quantity in each column.

Column A	Column B
$15 + 2\sqrt{50} - 15$	$15 - 15$
$2\sqrt{50}$	0

Step 3 It is now clear that column A is greater than column B, thus choice A is correct.

Exercises: SUMMARY DIRECTIONS FOR COMPARISON QUESTIONS

<u>Answer:</u>
- **A** if the quantity in Column A is greater;
- **B** if the quantity in Column B is greater;
- **C** if the two quantities are equal;
- **D** if the relationship cannot be determined from the information given.

AN "E" RESPONSE WILL NOT BE SCORED.

	Column A	Column B
1)	$\sqrt{2} + \sqrt{5}$	$\sqrt{7}$
2)	$\sqrt{10} + \sqrt{10}$	$\sqrt{20}$
3)	$\sqrt{22}$	$\sqrt{5} + \sqrt{17}$
4)	$\sqrt{36}$	$\sqrt{10} + \sqrt{26}$
5)	$\sqrt{5} + \sqrt{26}$	$\sqrt{31}$
6)	7	$\sqrt{3} + \sqrt{24}$
7)	9	$\sqrt{8} + \sqrt{13}$
8)	$\sqrt{10} + \sqrt{10}$	6
9)	$\sqrt{26}$	$\sqrt{2} + \sqrt{8}$
10)	$\sqrt{23} + \sqrt{3}$	$\sqrt{50}$

Algebra: Shortcuts 15 - 17

Shortcut #15 Solving for an algebraic expression given 1 equation

Multiply or divide both sides of the equation by a number that will make one side of the equation identical to the expression to be solved.

Example 1 **If $3x + 5y = 7$, then $6x + 10y = ?$**

Step 1 Multiply both sides of the equation, $3x + 5y = 7$, by 2 in order to obtain the expression $6x + 10y$.

Note: If one side of an equation is multiplied or divided by a number, then the other side of the equation **must** be multiplied or divided by that same number. Also, each term of the equation **must** be multiplied or divided by that same number.

$$2\,(3x + 5y = 7)$$

$$\mathbf{6x + 10y = 14}$$

Example 2 **If $3x + 6y = 12$, then $x + 2y = ?$**

Step 1 Divide both sides of the equation by 3 in order to obtain the expression $x + 2y$. Notice that each side of the equation and each term **must** be divided by 3 .

$$\frac{3x + 6y}{3} = \frac{12}{3}$$

$$\frac{3x}{3} + \frac{6y}{3} = \frac{12}{3}$$

$$\mathbf{x + 2y = 4}$$

Exercises: **Determine the numerical value of the expression.**

1) If $3x - 2y = 5$, then $9x - 6y = ?$

2) If $2x + y = -3$, then $8x + 4y = ?$

3) If $x + 2y = 3$, then $10x + 20y = ?$

4) If $5x + 10y = 4$, then $15x + 30y = ?$

5) If $x - y = 7$, then $3x - 3y = ?$

6) If $2x + 4y = 8$, then $x + 2y = ?$

7) If $8x - 10y = 2$, then $4x - 5y = ?$

8) If $5x + 15y = 20$, then $x + 3y = ?$

9) If $3x - 3y = 9$, then $x - y = ?$

Shortcut #16 Solving for an algebraic expression given 2 equations

1st) Add or subtract the equations in order to obtain a new equation. Adding the equations is usually the proper operation but this will depend on the problem. If the answer is not obtained at this point, perform the next step.

2nd) If necessary, multiply or divide this new equation by a particular number so that one side of the equation is identical to the expression to be solved.

Example 1 **If $2x + y = 5$ and $x + 3y = 6$, then $3x + 4y = ?$**

Step 1 First, attempt to obtain the expression $3x + 4y$ by adding the equations.

$$\begin{array}{rrcl} & 2x + y & = & 5 \\ + & x + 3y & = & 6 \\ \hline & \mathbf{3x + 4y} & = & \mathbf{11} \end{array}$$

Note: Adding the equations was the correct step because the new equation contains the expression $3x + 4y$. The new equation reveals the answer: $\mathbf{3x + 4y = 11}$.

Example 2 **If $2x - y = 9$ and $3x - 2y = -2$, then $10x - 6y = ?$**

Step 1 First attempt to obtain the expression $10x - 6y$, by adding the equations.

$$\begin{array}{rrcl} & 2x - y & = & 9 \\ + & 3x - 2y & = & -2 \\ \hline & 5x - 3y & = & 7 \end{array}$$

Step 2 The new equation does **not** contain the expression $10x - 6y$, however if both sides of the equation are multiplied by 2, the expression $10x - 6y$ and its' numerical value will be obtained. Remember to multiply each side of the equation and each term by 2.

$$2(5x - 3y = 7)$$

$$\mathbf{10x - 6y = 14}$$

Example 3 **If $2x + 3y = 7$ and $3x + 2y = 8$, then $x + y = ?$**

Step 1 First, attempt to obtain the expression $x + y$ by adding the equations.

$$\begin{array}{rrcl} & 2x + 3y & = & 7 \\ + & 3x + 2y & = & 8 \\ \hline & 5x + 5y & = & 15 \end{array}$$

Step 2 The new equation does **not** contain the expression $x + y$, however if both sides of the equation are divided by 5, the expression $x + y$ and its' numerical value will be obtained. Remember to divide each side of the equation and each term by 5.

$$\frac{5x + 5y}{5} = \frac{15}{5}$$

$$\frac{5x}{5} + \frac{5y}{5} = \frac{15}{5}$$

$$\boldsymbol{x + y = 3}$$

Example 4 **If $3x + 5y = 4$ and $2x + 2y = 3$, then $x + 3y = ?$**

Step 1 If the two equations are added, the expression $x + 3y$ will **not** be obtained even if the new equation is multiplied or divided by a particular number. Therefore, subtract the equations in order to obtain a new equation.

Note: When subtracting equations, it is usually best to place the first equation above the second equation, although this is not necessary.

$$\begin{array}{rrcl} & 3x + 5y & = & 4 \\ -(& 2x + 2y & = & 3\ ) \\ \hline & \mathbf{x + 3y} & = & \mathbf{1} \end{array}$$

Note: The new equation created by the subtraction of two equations may need to be multiplied or divided by a particular number so that the one side of the equation is identical to the expression to be solved.

Note: If the opposite of the desired expression is obtained, multiply the **entire** equation by -1 in order to obtain the correct expression. Expressions are opposites if each expression contains identical terms and the signs of the identical terms are opposites. An example of expressions that are opposites is: $x - 3y$ and $-x + 3y$.

Exercises: **Determine the numerical value of the given expression.**

1) If $3x + 2y = 3$ and $4x + y = 5$, then $7x + 3y = ?$

2) If $x + 2y = 1$ and $2x + y = 2$, then $3x + 3y = ?$

3) If $2x - y = 6$ and $x - y = 7$, then $3x - 2y = ?$

4) If $3x + y = 2$ and $x - 4y = 5$, then $8x - 6y = ?$

5) If $x - y = 3$ and $x - 2y = -4$, then $6x - 9y = ?$

6) If $2x + 3y = 10$ and $3x + 2y = 5$, then $x + y = ?$

7) If $5x - y = 8$ and $x - 5y = -2$, then $x - y = ?$

8) If $3x + 5y = 12$ and $2x + 3y = 5$, then $x + 2y = ?$

9) If $7x + 2y = 3$ and $2x - y = 4$, then $5x + 3y = ?$

Shortcut #17 Solving for a fraction given two proportions

From the two proportions, multiply the two fractions that contain variables (letters).
If the product equals the fraction to be solved, then the numeric value of this product is the answer.

Note: A **proportion** is an equation in which two fractions are set equal to each other.

Example *If* $\frac{a}{b} = \frac{2}{3}$ *and* $\frac{b}{c} = \frac{4}{5}$ *, then* $\frac{a}{c} = ?$

Step 1 Multiply: $\frac{a}{b} \times \frac{b}{c} = \frac{a}{c}$. Thus, the numerical value of $\frac{a}{b} \times \frac{b}{c}$ will equal the numerical value of $\frac{a}{c}$.

Note: Notice that the variable "b" cancels out in this multiplication.

$$\frac{a}{\cancel{b}} \times \frac{\cancel{b}}{c} = \frac{a}{c}$$

Step 2 Substitute the numerical values for $\frac{a}{b}$ and $\frac{b}{c}$ and multiply.

$$\frac{a}{b} \times \frac{b}{c} = \frac{a}{c}$$

$$\frac{2}{3} \times \frac{4}{5} = \mathbf{\frac{8}{15}} \qquad \textit{Thus, } \mathbf{\frac{a}{c}} = \mathbf{\frac{8}{15}}$$

Exercises: **Determine the numerical value of $\mathbf{\frac{a}{c}}$.**

1) $\frac{a}{b} = \frac{1}{2}$ *and* $\frac{b}{c} = \frac{3}{5}$ **2)** $\frac{a}{b} = \frac{3}{4}$ *and* $\frac{b}{c} = \frac{9}{7}$

3) $\frac{a}{b} = \frac{2}{3}$ *and* $\frac{b}{c} = \frac{4}{5}$ **4)** $\frac{a}{b} = \frac{3}{2}$ *and* $\frac{b}{c} = \frac{3}{8}$

Geometry: Shortcuts 18 - 24

Shortcut #18 Calculating the area formed by two overlapping shapes

The area formed by two overlapping shapes can be calculated by subtracting the area of the larger shape by the area of the smaller shape. This procedure applies only if one shape is completely within another shape.

Example 1 **The following circles have the same center (also known as concentric circles). The radius of the larger circle is 7 and the radius of the smaller circle is 3. Determine the area of the shaded region.**

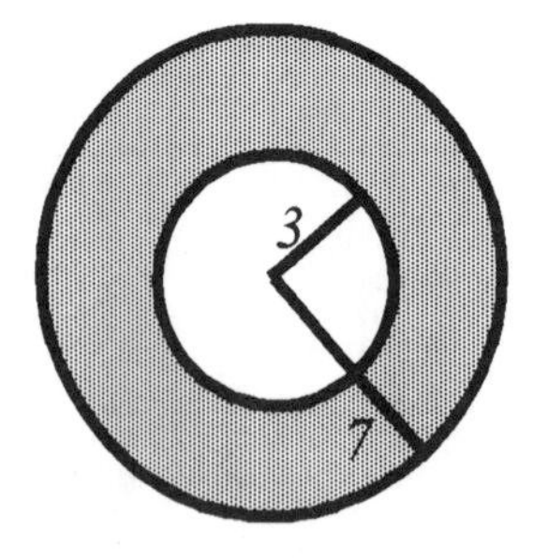

Step 1 One circle is completely within the other circle, therefore the area of the shaded region will equal the area of the larger circle subtracted by the area of the smaller circle. Determine the area of each circle. The area formula for a circle is: $\mathbf{A = \pi r^2}$, where A = area , r = radius and π (pi) is approximately 3.14 or 22/7 .

Note: If the exact area of a circle is required, do **not** substitute a numerical value for π. This is because the exact value for π can **not** be represented numerically.

Note: Recall that the absence of a math operation indicates multiplication, thus πr^2 represents π multiplied by r^2. The correct **order of operations** is to perform the exponent operation before the multiplication. Thus, the radius (r) **must** be squared **before** it is multiplied by π.

Note: π is a Greek letter and its' exact value can only be expressed by this letter. The algebraic rules that apply to variables (letters) also apply to π. For example, $6\pi - 2\pi = 4\pi$ and $(\pi)(\pi) = \pi^2$.

Area of larger Circle:

$A = \pi r^2$

$A = \pi(7)^2$

$A = \pi 49$ or 49π

Area of smaller Circle:

$A = \pi r^2$

$A = \pi(3)^2$

$A = \pi 9$ or 9π

Step 2 The area of the shaded region will equal the area of the larger circle subtracted by the area of the smaller circle.

Area of shaded region = (Area of larger Circle) – (Area of smaller Circle)

Area of shaded region = $49\pi - 9\pi = 40\pi$

Example 2 A circle with center F (also known as circle F) is inscribed in square ABCD. Square ABCD has length 6. Determine the area of the shaded region.

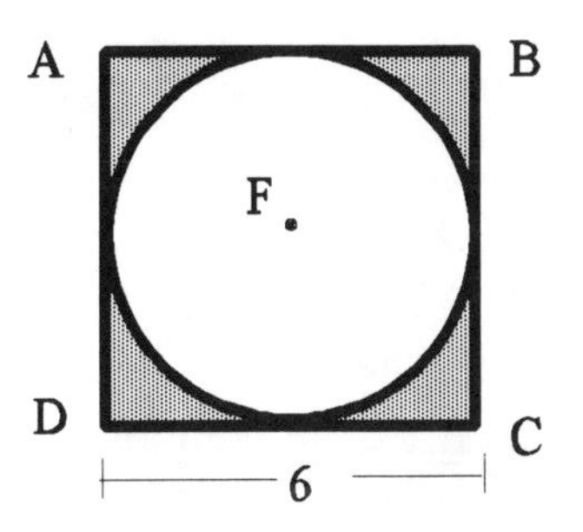

Note: A circle is **inscribed** in a square if the circle is inside the square and each side of the square is **tangent** (touches at one point) to the circle.

Step 1 The circle is completely within the square, therefore the area of the shaded region will equal the area of the square subtracted by the area of the circle. Determine the area of the square and the circle. The area formula of a square is: $\mathbf{A = Lw}$, where A = area, L = length and w = width. Notice that in a square the length equals the width. The area formula of a circle is $\mathbf{A = \pi r^2}$, where A = area, π = pi and r = radius.

Note: The radius of the circle equals half the side of the square, therefore the radius is 3 .

Area of the Square:

$A = Lw$

$A = (6)(6)$

$A = 36$

Area of the Circle:

$A = \pi r^2$

$A = \pi(3)^2$

$A = \pi 9$ or 9π

Step 2 The area of the shaded region will equal the area of the square subtracted by the area of the circle.

Area of shaded region = (Area of the Square) – (Area of the Circle)

Area of shaded region = 36 – 9π

Note: This answer can **not** be simplified any further because 36 and 9π are unlike terms.

Example 3 Square ABCD has length 10. The circle has a center at point D and a radius of 5. Determine the area of the shaded region.

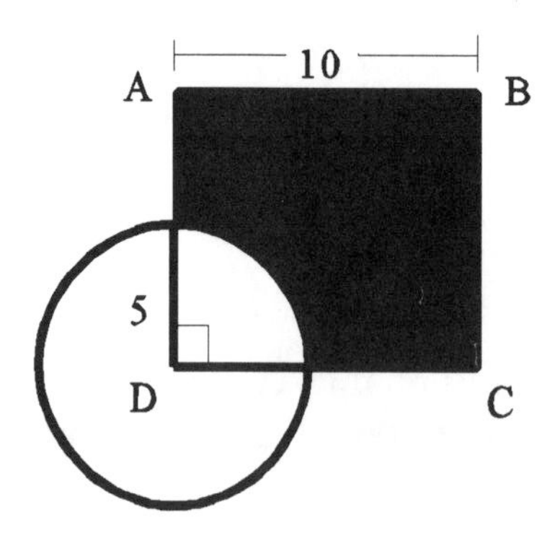

Note: A 90° angle (also known as a right angle) is indicated by a small square (⊾).

Step 1 Consider the two overlapping shapes to be the square and one-fourth of the circle. One-fourth of the circle overlaps the square because the circle's central angle of 90° overlaps the sides of the square. Recall that the sum of the angles around the center of a circle is 360, thus a sector that contains a central angle of 90° represents one-fourth the area of a circle because 90°/360° equals ¼. The **central angle of a circle** is an angle whose vertex (common endpoint) is the center of a circle. A **sector of a circle** is the region bounded by two radii and an arc of a circle.

Step 2 The area of the shaded region will equal the area of the square subtracted by one-fourth the area of the circle. Determine the area of the square and one-fourth the area of the circle.

Area of the Square:

$A = Lw$

$A = (10)(10)$

¼ of the Area of the Circle:

$A = \frac{1}{4}\pi r^2$

$A = \frac{1}{4}\pi(5)^2$

$A = 100$ $\qquad\qquad$ $A = \dfrac{25\pi}{4}$

Step 3 The area of the shaded region will equal the area of the square subtracted by one-fourth the area of the circle.

Area of shaded region = (Area of the Square) – (¼ Area of the Circle)

Area of shaded region = $100 - \dfrac{25\pi}{4}$

Example 4 A circle with center A has a radius of 3. AC ⊥ AB. Determine the area of the shaded region.

Note: **Perpendicular lines** form a 90° angle (right angle). A right angle is indicated by the symbol, ⊾. A triangle that contains a right angle is a **right triangle**.

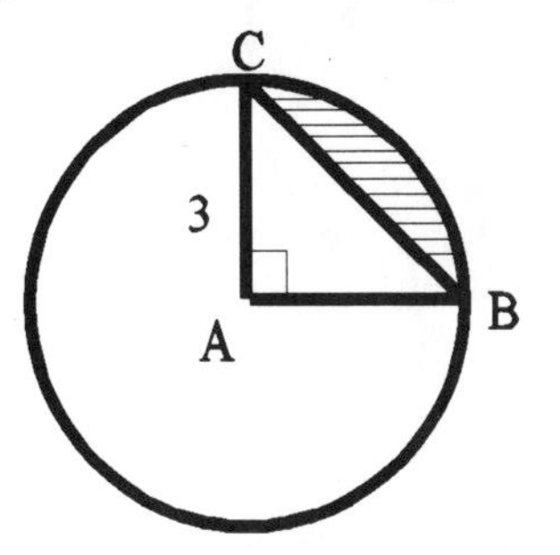

Step 1 Consider the two overlapping shapes to be the right triangle and one-fourth of the circle represented by sector CAB.

Step 2 The right triangle is completely within **one-fourth** of the circle, therefore the area of the shaded region will equal the area of one-fourth of the circle subtracted by the area of the triangle. Determine the area of the triangle and one-fourth of the circle. The area formula of a triangle is $A = ½ (b) (h)$, where b = base and h = height. The area formula of one-fourth of a circle is: $A = \dfrac{\pi r^2}{4}$ where r = radius, and π = pi (approximately 3.14 or 22/7).

Note: The radius of the circle is 3, therefore both AB and AC equal 3 because all radii of the same circle are equal. Either leg (AC or AB) of the right triangle can be considered the base or the height because they are perpendicular to one another. Thus, the base and height in the area formula of the triangle equal 3.

¼ of the Area of the Circle:

$$A = \frac{\pi r^2}{4}$$

$$A = \frac{\pi (3)^2}{4}$$

$$A = \frac{\pi (9)}{4} \text{ or } \frac{9\pi}{4}$$

Area of the Right Triangle:

$$A = ½ (b)(h)$$

$$A = ½ (3)(3)$$

$$A = ½ (9) = \frac{9}{2}$$

Step 3 The area of the shaded region will equal the area of one-fourth of the circle subtracted by the area of the triangle. Notice that the answer may be simplified into one fraction by finding a common denominator.

Area of shaded region = (¼ Area of the Circle) – (Area of the Triangle)

Area of shaded region = $\dfrac{9\pi}{4} - \dfrac{9}{2}$ **or** $\dfrac{9\pi - 18}{4}$

Exercises: **Find the area of the shaded region. The center of all the circles is point D.**

1)

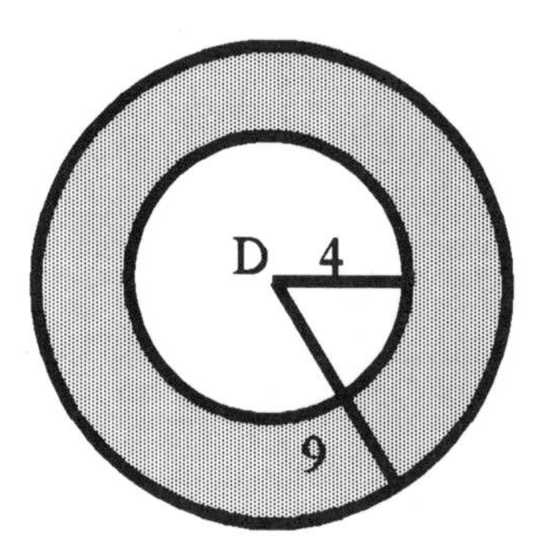

2)

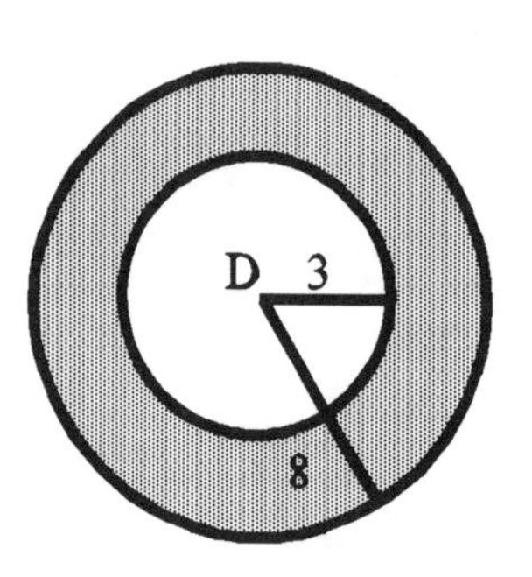

3) Both figures are rectangles.

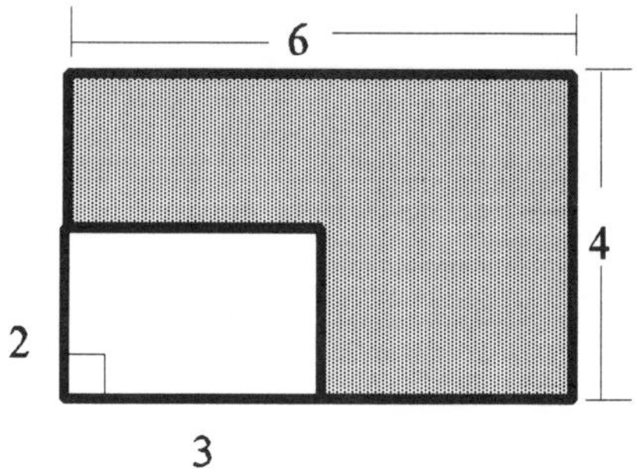

4) EFGH is a square

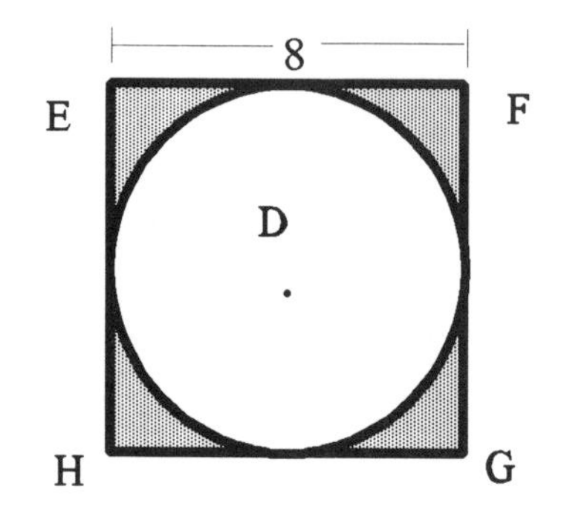

5) EFGH is a square

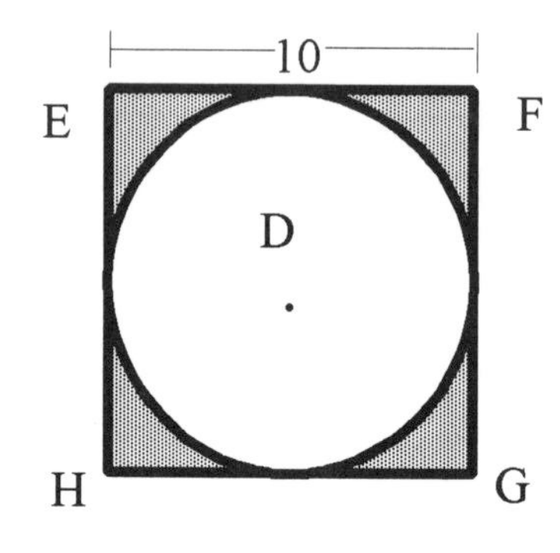

6) ABCD is a square.
Circle D has a radius of 4.

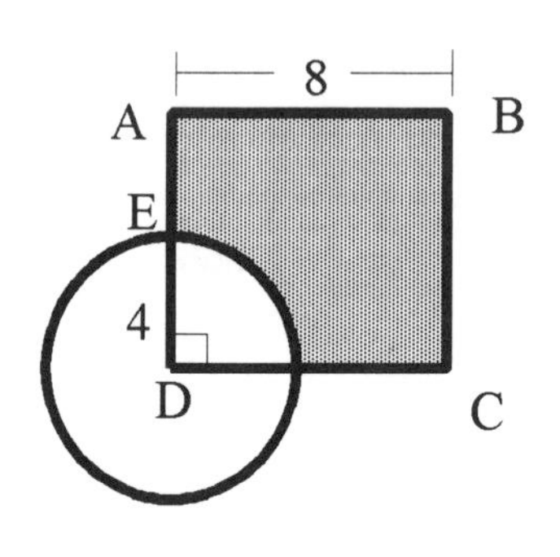

7) EFGH is a square

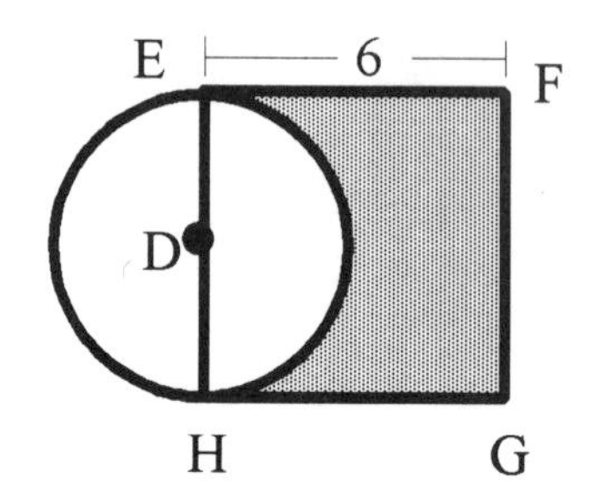

8)

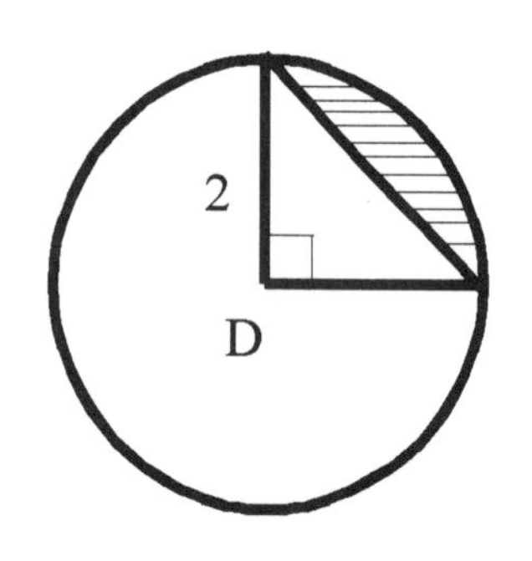

9)

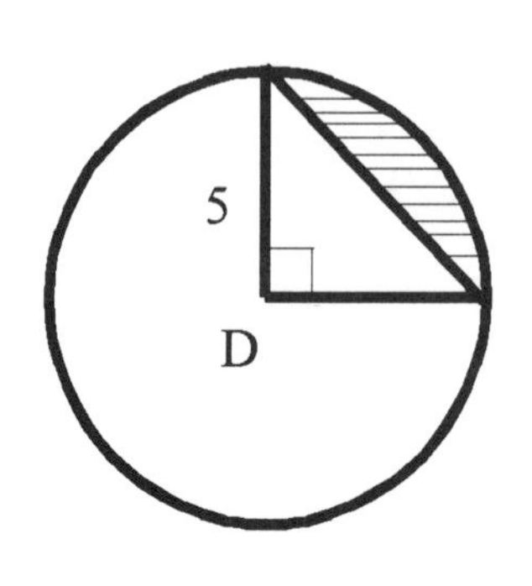

Shortcut #19 Comparing the sides of a triangle to their opposite angles

In any triangle, the longest side is opposite the greatest angle and the shortest side is opposite the smallest angle. Sides that are opposite equal angles are equal. Conversely, angles that are opposite equal sides are equal.

Example 1 **Which is greatest: a, b, or c ?**

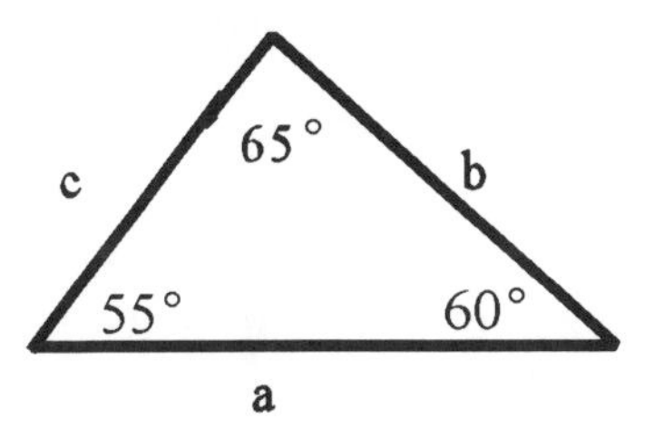

Step 1 "a" is greatest because it is the side opposite the greatest angle (65°).

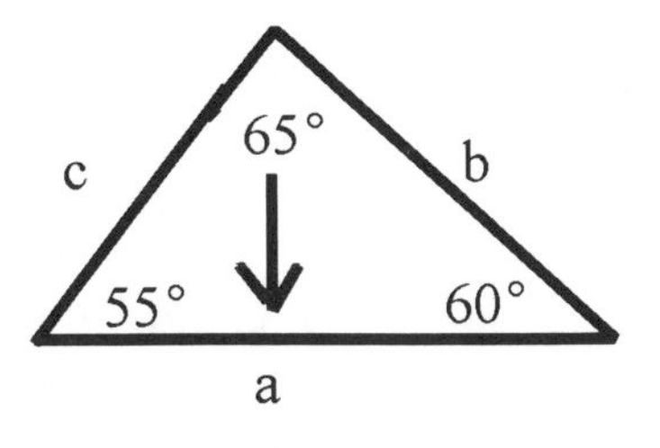

Example 2 **Which side is greater: x or y ?**

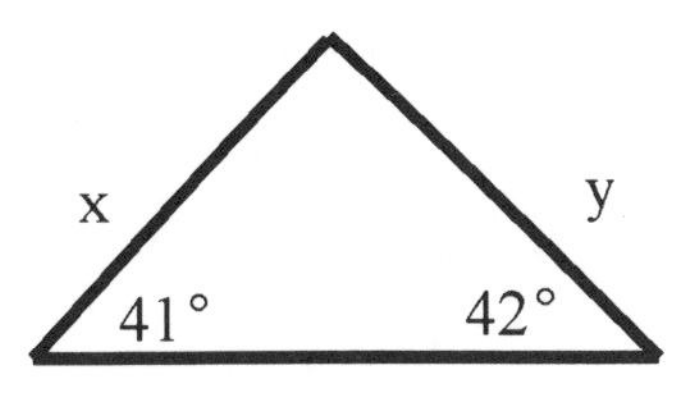

Step 1 "x" is greater than y because x is opposite a greater angle than y is opposite.
"x" is opposite the 42° angle and y is opposite the 41° angle.
Notice that the unlabeled angle does **not** affect the answer.

Example 3 **Which is greater: x or y ?**

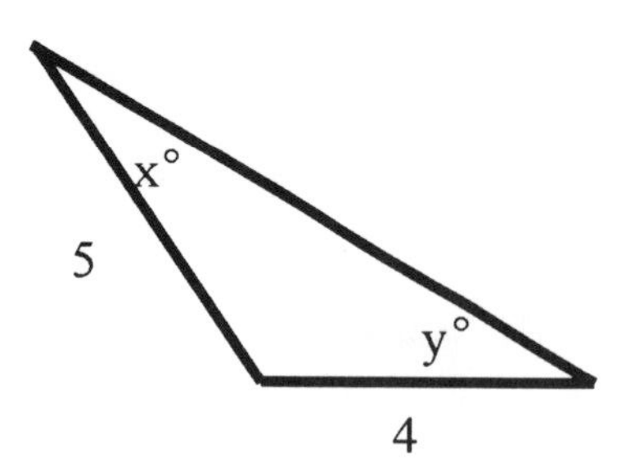

Step 1 "y" is greater than x because y is opposite a longer side than x is opposite. "y" is opposite the side of length 5 and x is opposite the side of length 4. Notice that the unlabeled side does **not** affect the answer.

Example 4 **List the angles of the triangle in descending order of measure.**

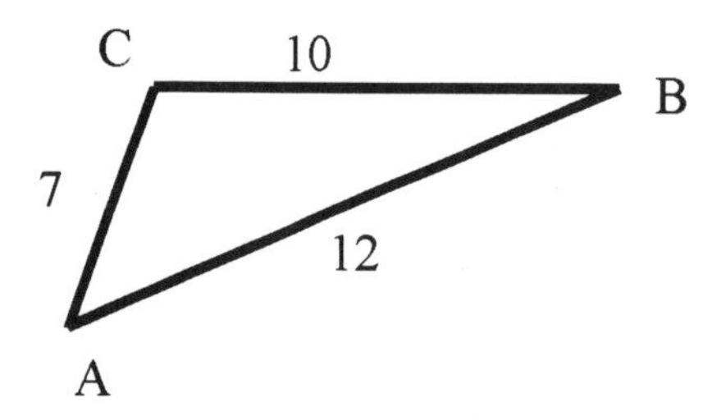

Step 1 $\angle C$ is the greatest angle because it is opposite the longest side (12). $\angle B$ is the smallest angle because it is opposite the shortest side (7). Thus, the descending order of measure of the angles is: $\angle \mathbf{C} > \angle \mathbf{A} > \angle \mathbf{B}$.

Example 5 **Determine the value of x .**

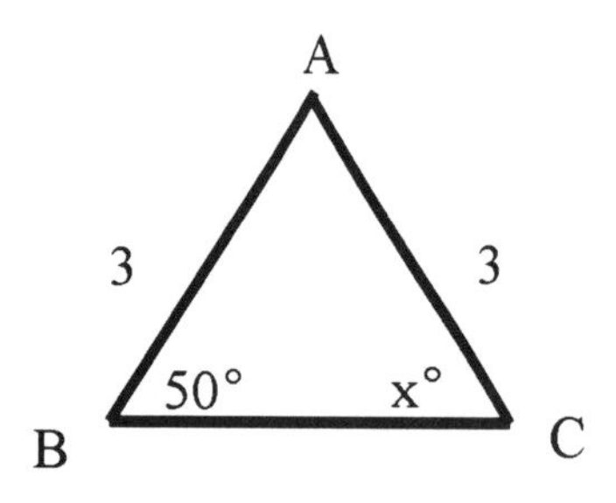

Step 1 $\angle B$ is opposite side AC and $\angle C$ is opposite side AB. $\angle B = \angle C$ because $\angle B$ and $\angle C$ are opposite equal sides. Therefore, $\angle B = \angle C = 50°$ and so $\mathbf{x = 50}$.

Example 6 **Determine the value of y .**

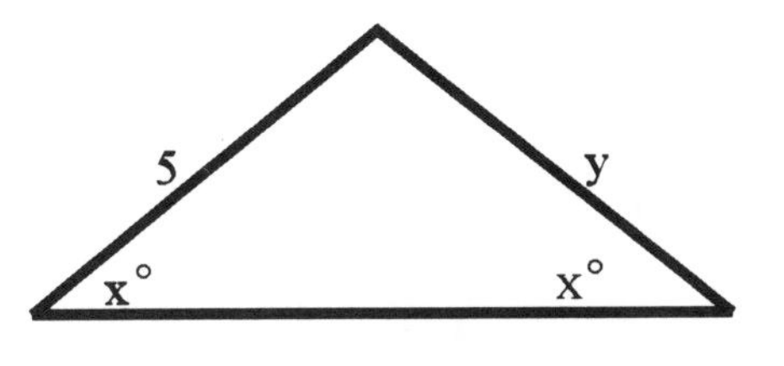

Step 1 Because "y" and the side of length 5 are both opposite angles of the same measure (x), these sides **must** be equal. Therefore, **y = 5.**

Exercises:

Which is greatest: x, y or z ?

1)

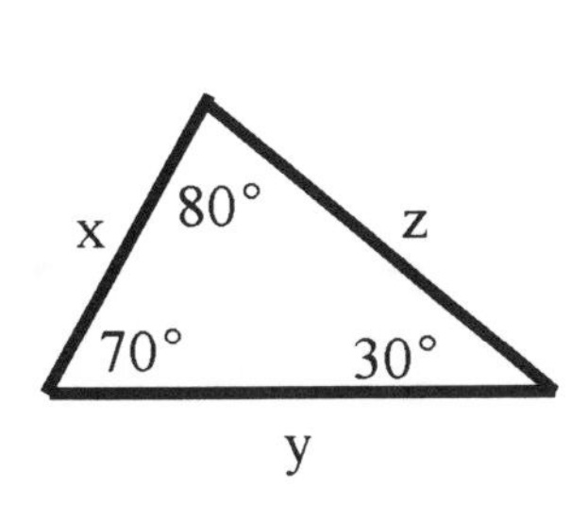

2)

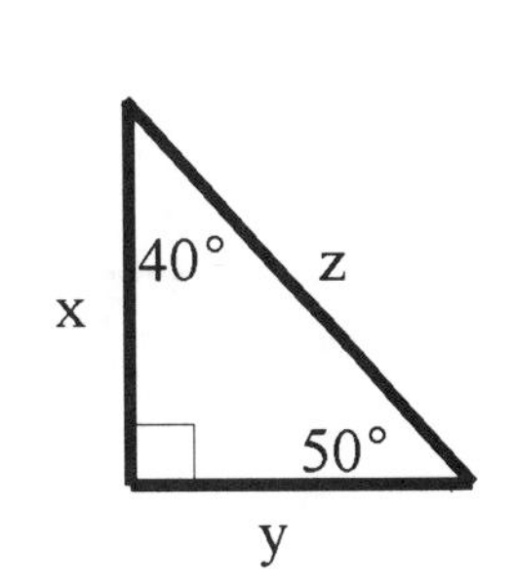

Which is greater: x or y ?

3)

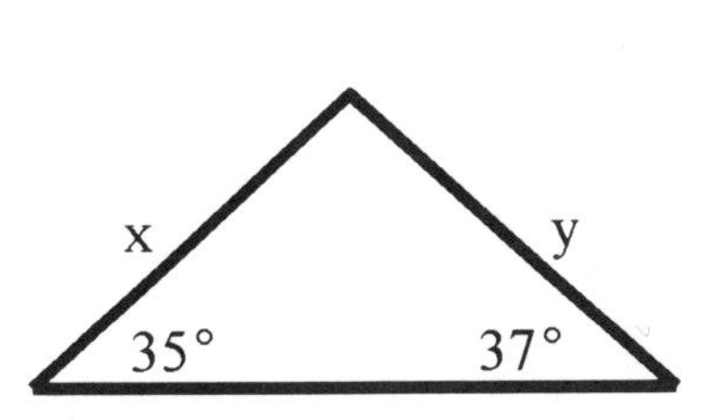

4)

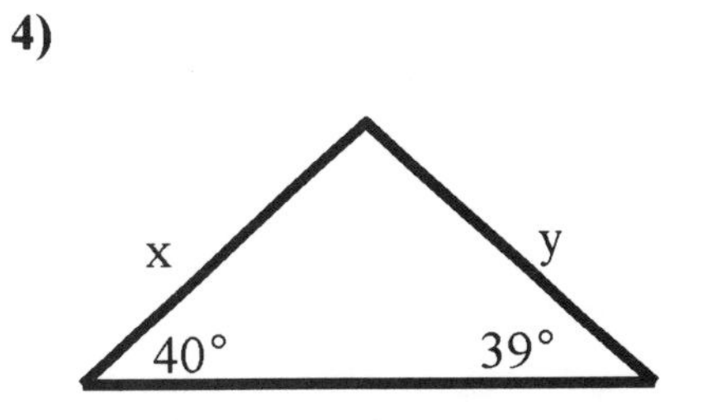

5)

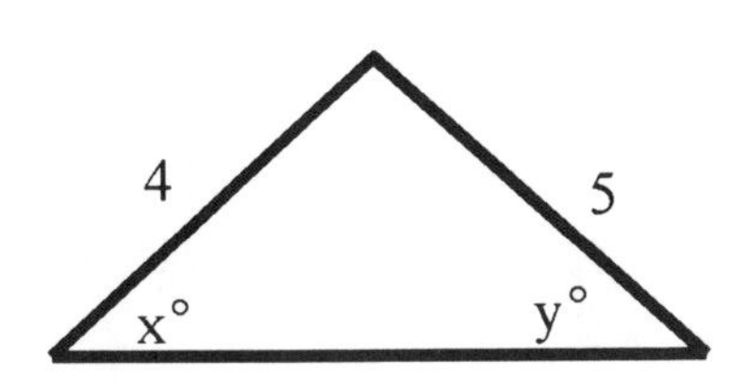

6)

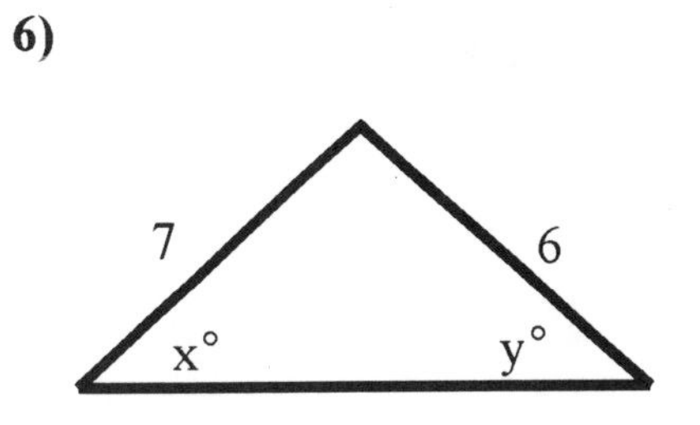

List the angles in descending order of measure.

7) **8)**

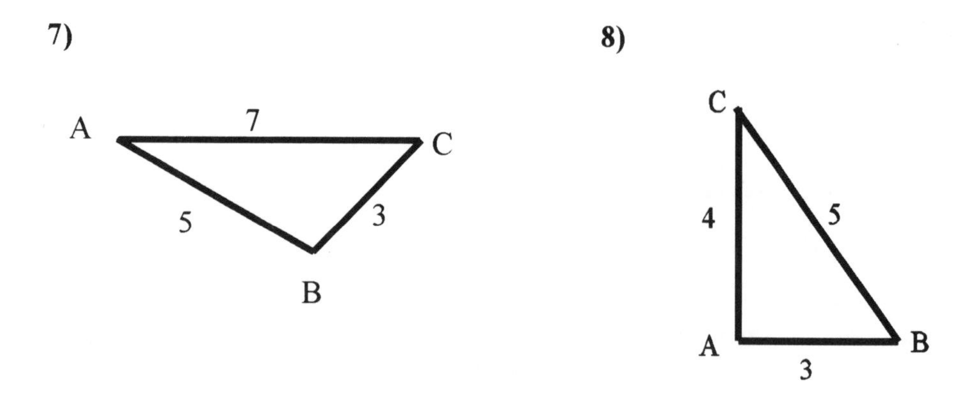

Determine the value of x .

9)

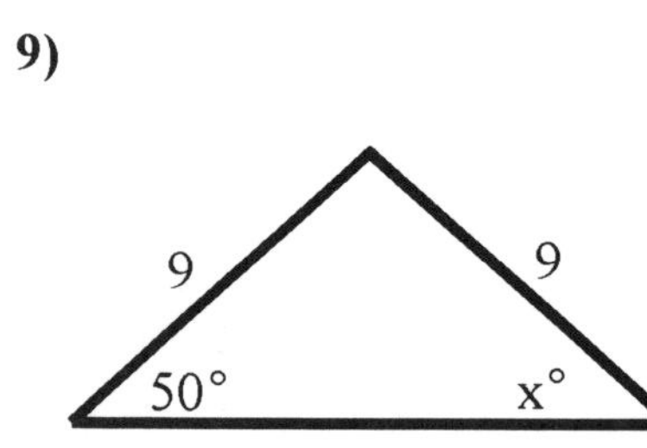

10)

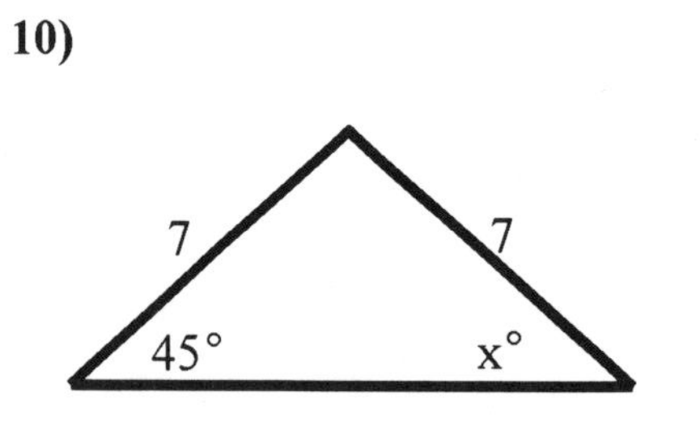

11)

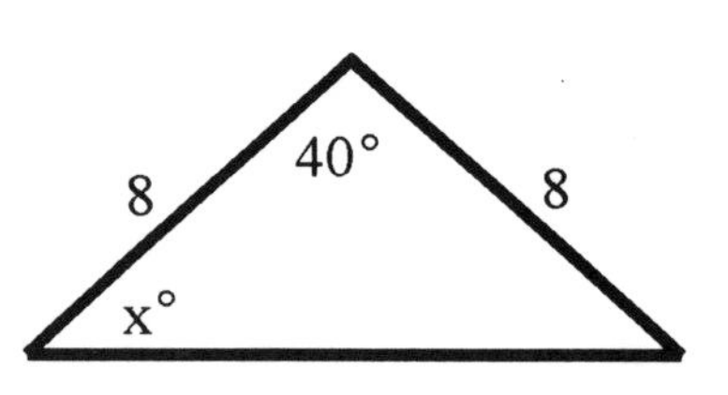

12)

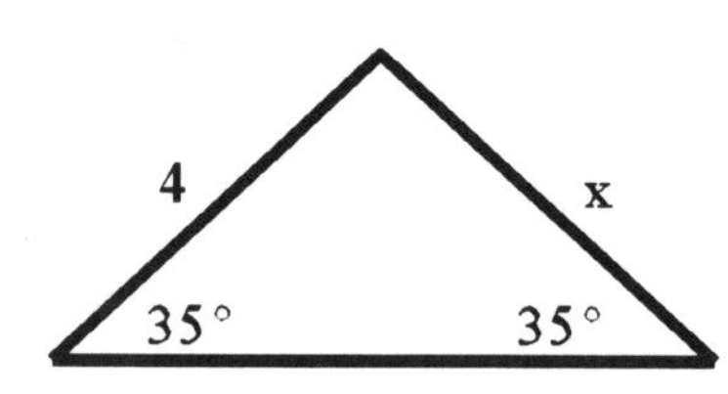

13)

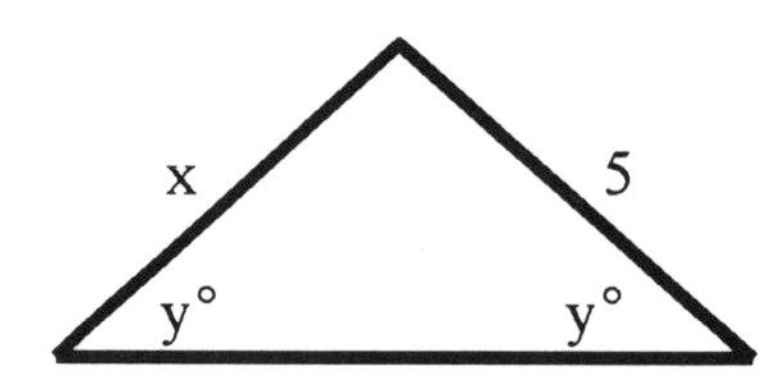

Shortcut #20 Determining the possible lengths of a missing side of a triangle

The length of any side of a triangle is <u>less</u> than the <u>sum</u> of the lengths of the other two sides of that triangle. Also, the length of any side of a triangle is <u>greater</u> than the <u>difference</u> in length of the other two sides of that triangle.

Example **What are the possible lengths of a side of a triangle, if the other two sides of the triangle have lengths 5 and 7? Represent the possible lengths as x .**

Step 1 The **greatest** possible length of the unknown side of a triangle is slightly **less** than the sum of the lengths of the two known sides ($5 + 7 = 12$). Therefore, the **greatest** possible length of the unknown side is just under 12 ($x < 12$). Notice that values such as 11 ¾ or 11.9 are possible values of x. However, 12 is **not** a possible value of x .

Step 2 The **least** possible length of the unknown side is slightly **greater** than the **difference** in length of the two known sides ($7 - 5 = 2$). Therefore, the **least** possible value of x is slightly greater than 2 ($2 < x$). Notice that values such as 2 ¼ or 2.1 are possible values of x , however, 2 is **not** a possible value of x .

Step 3 Combine the inequalities: $2 < x$ and $x < 12$.

$$2 < x < 12$$

This is the answer because it describes all possible values of x . Notice that the inequality signs in the combined inequality **must** point in the same direction and x **must** be in the middle. The answer represents all values between 2 and 12 but **not** including 2 or 12. The answer could also have been written as $12 > x > 2$.

Exercises: **What are the possible lengths of an unknown side of a triangle, x , if the other two sides are the following:**

1) 2 and 5 **2)** 2 and 6 **3)** 7 and 16 **4)** 5 and 13

5) 7 and 8 **6)** 1 and 3 **7)** 2 and 4 **8)** 10 and 10

Shortcut #21 When a triangle can be formed from three given sides

A triangle can be formed from three sides only if the longest side is less than the sum of the other two sides.

Example 1 Could the lengths 6, 10, 17 form a triangle?

Step 1 If the longest side is **less** than the **sum** of the other two sides, then a triangle can be formed. Because the longest side (17) is **not** less than the **sum** of the shorter two sides (6 + 10 = 16), the three sides 6, 10, 17 can **not** form a triangle.

Longest side:		Sum of shortest two sides:
17		6 + 10 = 16
17	>	**16**

Example 2 Could the lengths 3, 4, 6 form a triangle?

Step 1 If the longest of three sides is **less** than the **sum** of the other two sides, then a triangle could be formed. Because the longest side (6) is **less** than the **sum** of the shorter two sides (3 + 4 = 7), the three sides 3, 4, 6 **can** form a triangle.

Longest side:		Sum of shortest two sides:
6		3 + 4 = 7
6	<	7

Example 3 Could the lengths 3, 5, 8 form a triangle?

Step 1 If the longest of three sides is **less** than the **sum** of the shorter two sides, then a triangle **can** be formed. Because the longest side (8) is **not** less than the **sum** of the shorter two sides (3 + 5 = 8), the three sides can **not** form a triangle.

Longest side:		Sum of shorter two sides:
8		3 + 5 = 8
8	=	**8**

Exercises: Could the following lengths form a triangle?

1) 3, 5, 7 **2)** 3, 5, 9 **3)** 7, 9, 12 **4)** 3, 3, 6

5) 3, 3, 3 **6)** 3, 4, 5 **7)** 5, 8, 10 **8)** 1, 1, 5

Shortcut #22 Each radius of the same circle is equal in length

The radius of a circle is a line in which one endpoint is the center of the circle and the other endpoint is on the circle. Recognizing that radii (plural for radius) of the same circle are equal is often the first key step in solving problems involving circles.

Example 1 Point "A" is the center of the circle. Solve for x.

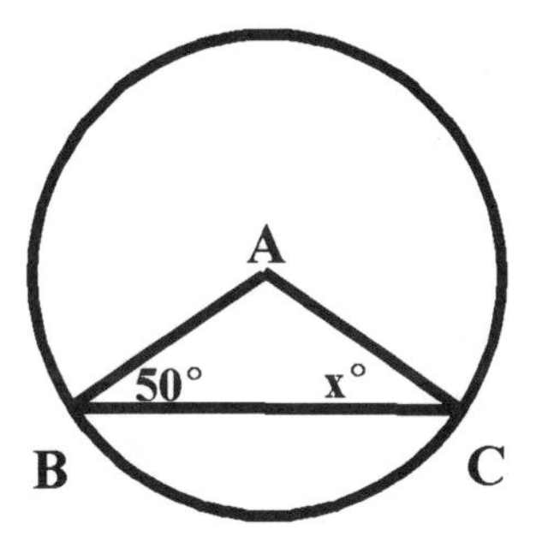

Step 1 AB and AC are both radii (plural for radius) of the circle. AB = AC because each radius of the same circle is equal in length. Thus, triangle ABC is an **isosceles triangle** because two sides of the triangle are equal.

Step 2 $\angle B = \angle C$ because both angles are opposite equal sides (also described as base angles of an isosceles triangle are equal). $\angle B$ is opposite side AC and $\angle C$ is opposite side AB. Because $\angle B = \angle C$, **x = 50**.

Example 2 Point "A" is the center of the circle. Solve for x.

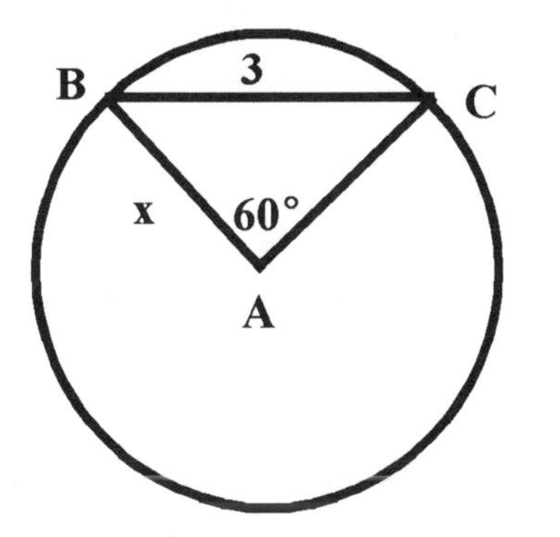

Step 1 AB = AC because each radius of the same circle is equal in length.

Step 2 $\angle B = \angle C$ because both angles are opposite equal sides ($\angle B$ is opposite side AC and $\angle C$ is opposite side AB).

Step 3 $\angle B$ and $\angle C$ both equal 60° because: 1) $\angle B = \angle C$ and 2) the sum of all three angles of any triangle is 180°.

$$\angle A + \angle B + \angle C = 180°$$

$$60° + 60° + 60° = 180°$$

Step 4 Triangle ABC is an **equilateral triangle** (all three sides are equal) because all three angles are equal. Therefore, **x = 3**.

Example 3 Point "A" is the center of the circle. $AB \perp CD$. CD = 10. Find the length of BC.

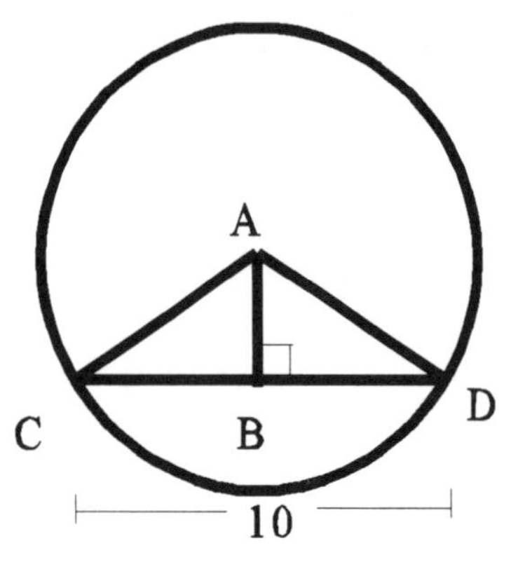

Step 1 AC = AD because each radius of the same circle is equal in length. Thus, triangle ACD is an **isosceles triangle** because two sides of the triangle are equal (AC = AD).

Note: 1) ∠CAD is the **vertex angle** of isosceles triangle ACD because ∠CAD is formed by two equal sides (AC = AD).

2) CD is the **base** of isosceles triangle ACD because side CD is opposite the vertex angle.

3) AB is the **height** from the vertex angle (∠CAD) to the base (CD).

Step 2 The height of an isosceles triangle from its' vertex angle, bisects the base of the isosceles triangle. Therefore, the height (AB) bisects the base (CD). Because CD is bisected at point B, BC = BD and so **BC = 5**.

Example 4 Point "A" is the center of the circle. ∠A = 90° and BC = $5\sqrt{2}$. Solve for x.

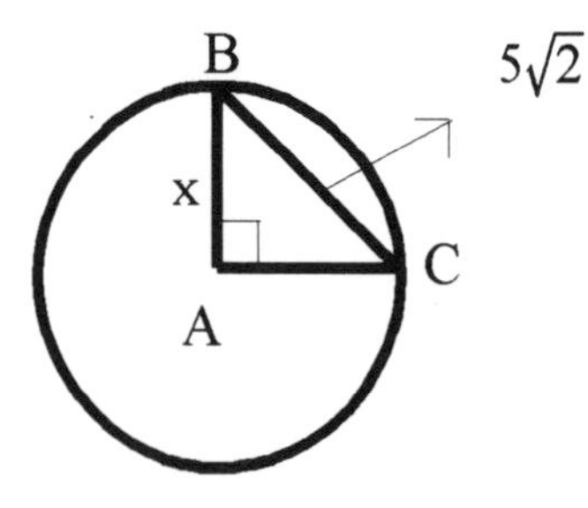

Step 1 AB = AC because each radius of the same circle is equal in length. Therefore, triangle ABC is an **isosceles right triangle** because two sides of the triangle are equal and the triangle contains a right angle (∠BAC).

Step 2 Solve for x by using the Pythagorean Theorem.

Note: The **Pythagorean Theorem** is a formula used to find a missing side of a right triangle. The formula is $\mathbf{a^2 + b^2 = c^2}$, where "a" and "b" are the length of the legs and "c" is the length of the hypotenuse of the right triangle. The **hypotenuse** is the longest side of a right triangle and is the side opposite the right angle . The **legs of a right triangle** are the shorter two sides of a right triangle and are the sides that form the right angle.

Note: Because the legs of an **isosceles right triangle** are equal in length, the same variable can be used for "a" and "b" in the Pythagorean Theorem. Here, "x" will represent the length of each leg of the isosceles right triangle (AB and AC).

$$a^2 + b^2 = c^2$$

$$x^2 + x^2 = c^2$$

$$2x^2 = (5\sqrt{2})^2$$

$$2x^2 = (25)(\sqrt{4})$$

$$2x^2 = (25)(2)$$

$$2x^2 = 50$$

$$x^2 = 25$$

$$\mathbf{x = 5}$$

Example 5 **Point "D" is the center of the circle and the radius of the circle is 2. The circle is tangent to the rectangle at points A, B and C. Side EF = 5. Determine the area of the rectangle.**

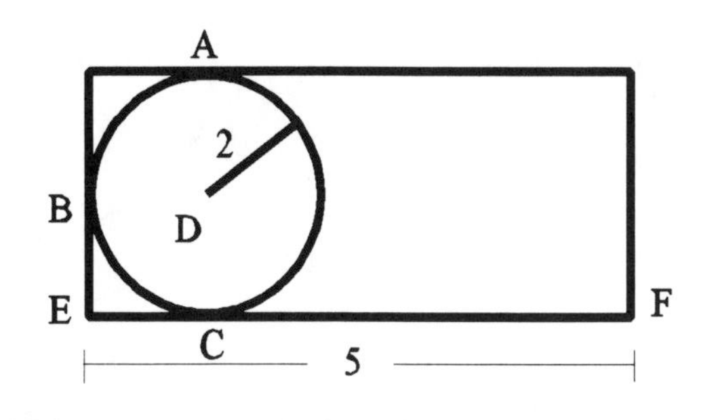

Step 1 The area formula of a rectangle is: A = LW, where L = length and W = width. Draw the circle's diameter, AC. Notice that the length of side AC is equal to the width of the rectangle. The length of the diameter (AC) is 4 because it is twice the length of the radius. Thus, the width of the rectangle is 4.

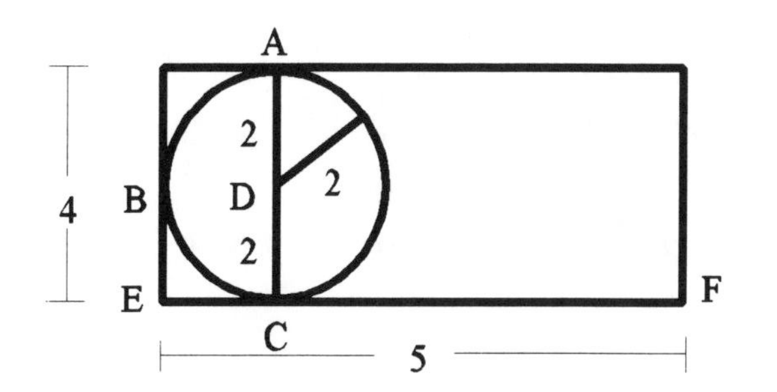

Step 2 The area formula of a rectangle can now be used.

$$A = LW$$

$$A = (5)(4)$$

$$\mathbf{A = 20}$$

Exercises: Point A is the center of the circles. Solve for x.

1)

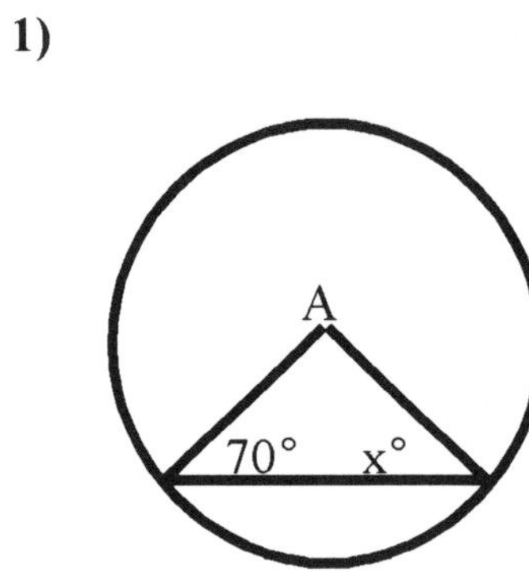

2)

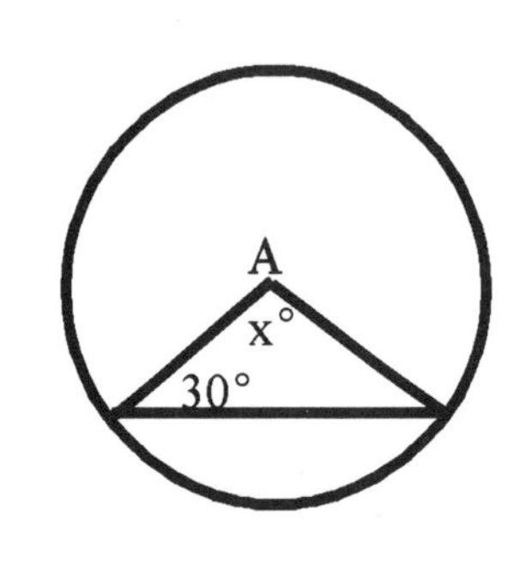

3)

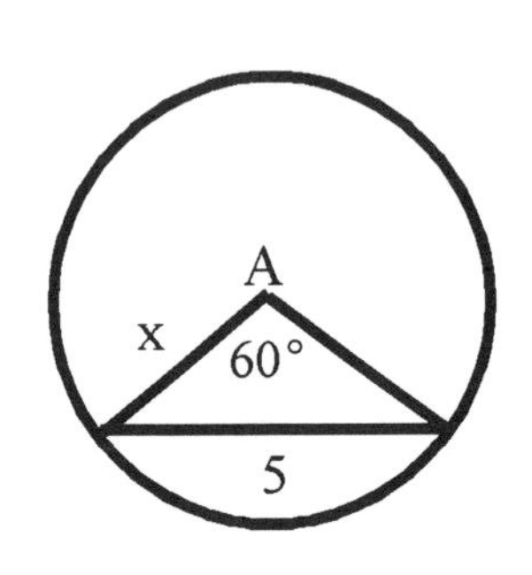

4) BC = 12

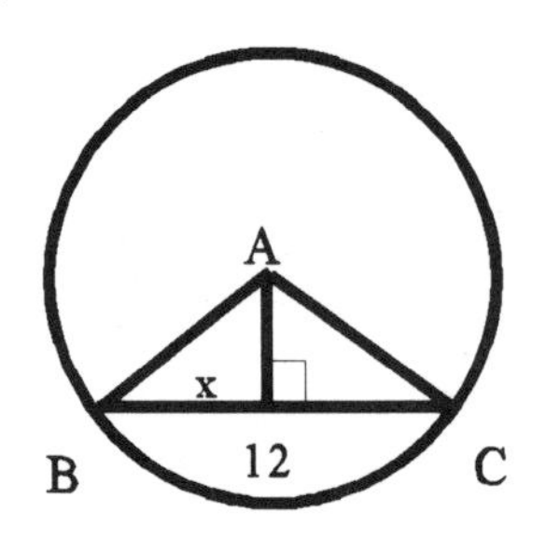

5) BC = 8

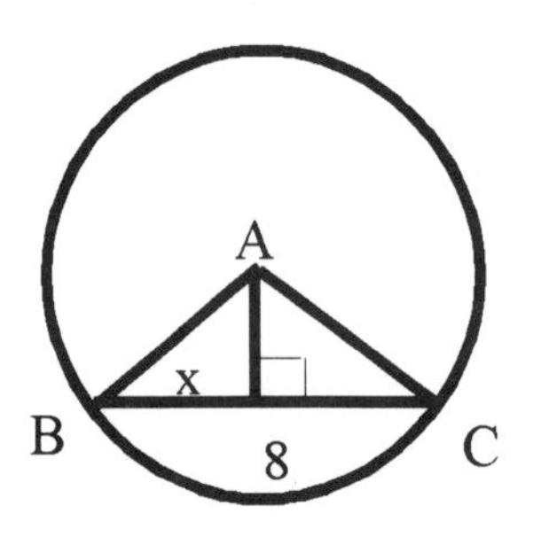

6) BC = $3\sqrt{2}$

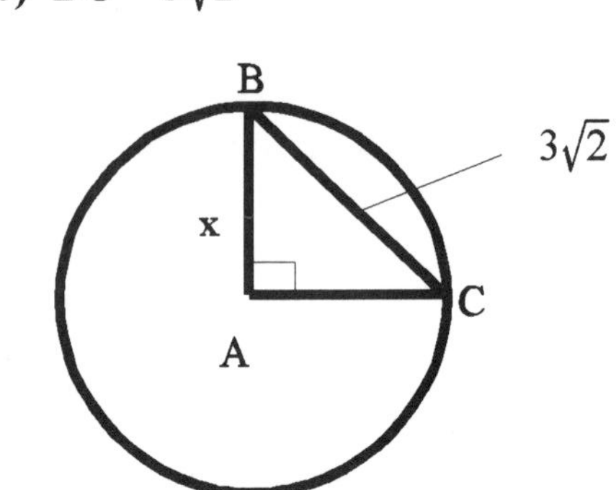

7) BC = $\sqrt{2}$

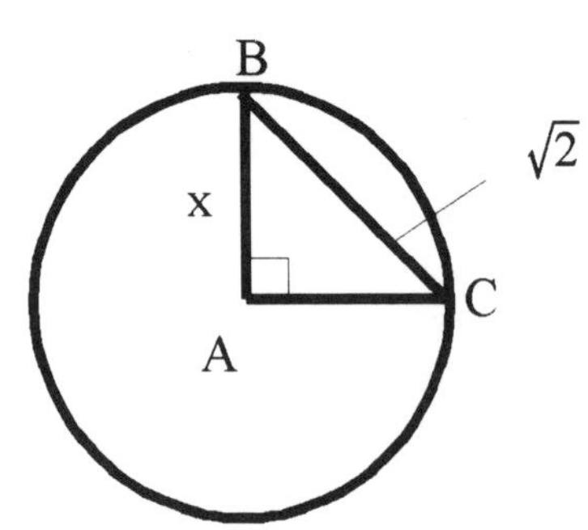

In problems 8 and 9, point "A" is the center of the circle. Points B, C and D are tangent points. Determine the area of the rectangle.

8) **9)**

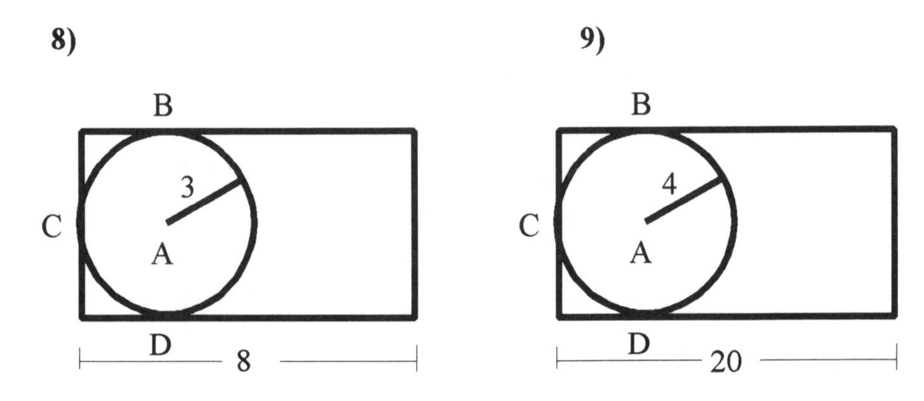

Point "A" is the center of the circle. The circle is inscribed in the square. Determine the area of the square.

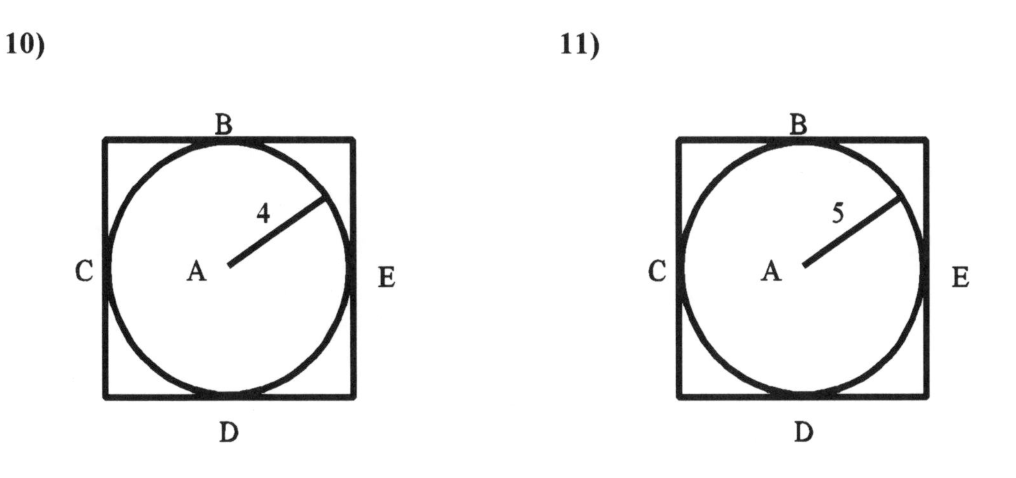

Shortcut #23 Pythagorean Triples

If any two lengths of a right triangle are known, then the length of the missing side may <u>always</u> be determined by using the Pythagorean Theorem*. However, a quicker method known as <u>Pythagorean Triples</u> may <u>sometimes</u> be used to find the length of the missing side of a right triangle. The two most common Pythagorean Triples are: 3, 4, 5 and 5, 12, 13. These two triples should be memorized.

If any two numbers from a Pythagorean Triple are lengths of a right triangle and the greatest number from that Pythagorean Triple corresponds to or will correspond to the hypotenuse, then the length of the missing side will be the number missing from that Pythagorean Triple. The greatest number from a Pythagorean Triple must correspond to the hypotenuse because the hypotenuse is always the longest side of a right triangle.

Note: A **right triangle** is a triangle that contains a 90° angle (right angle). A **right angle** is indicated by a small square (∟).

Note: The **hypotenuse** is the side of a right triangle that is opposite the 90° angle (right angle) and is always the longest side of a right triangle. The shorter two sides of a right triangle form a right angle and are known as the **legs.**

* The **Pythagorean Theorem** is $\mathbf{a^2 + b^2 = c^2}$, where "a" and "b" represent the two legs of the right triangle and "c" represents the hypotenuse of the right triangle.

Additional Pythagorean Triples may be created by multiplying all three numbers of a Pythagorean Triple by the same number. The following are examples of two Pythagorean Triples created by multiplying all three numbers of the Pythagorean Triple 3, 4, 5 by the same number.

3, 4, 5 multiplied by 2 creates the Pythagorean Triple 6, 8, 10:

3, 4, 5 multiplied by 3 creates the Pythagorean Triple 9, 12, 15:

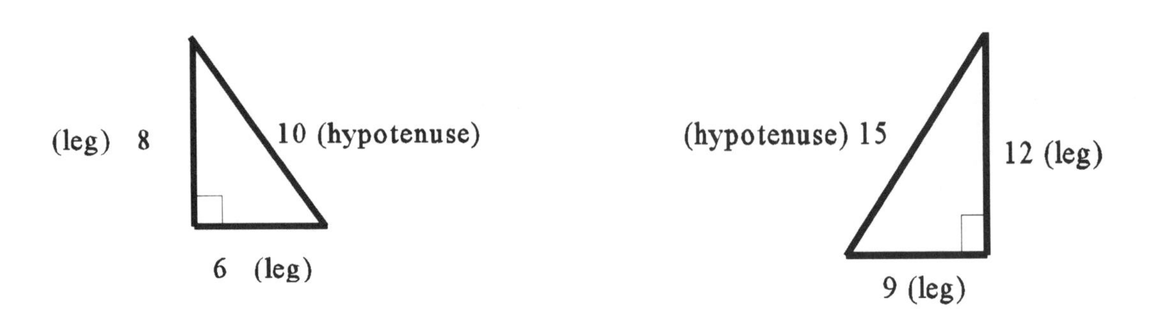

Example 1 Solve for x. Use a Pythagorean Triple if possible.

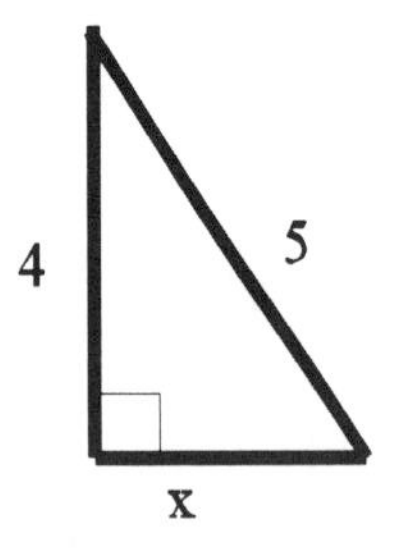

Step 1 The Pythagorean Triple 3, 4, 5 applies to this problem because: 1) two of the numbers from the Pythagorean Triple 3, 4, 5 are sides of the right triangle and 2) the greatest number, 5, in the Pythagorean Triple 3, 4, 5 corresponds to the hypotenuse.

Step 2 $\mathbf{x = 3}$ because 3 is the unused number from the Pythagorean Triple 3, 4, 5.

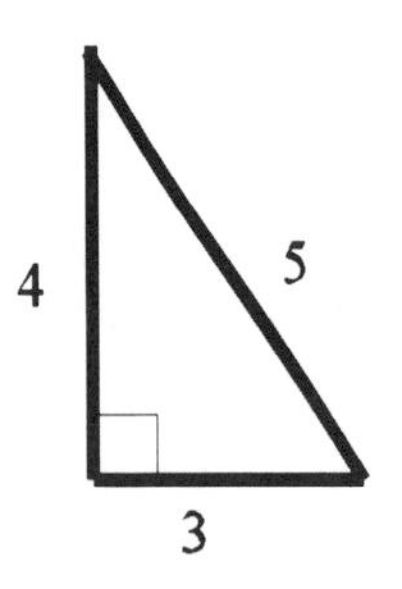

Example 2 Solve for x. Use a Pythagorean Triple if possible.

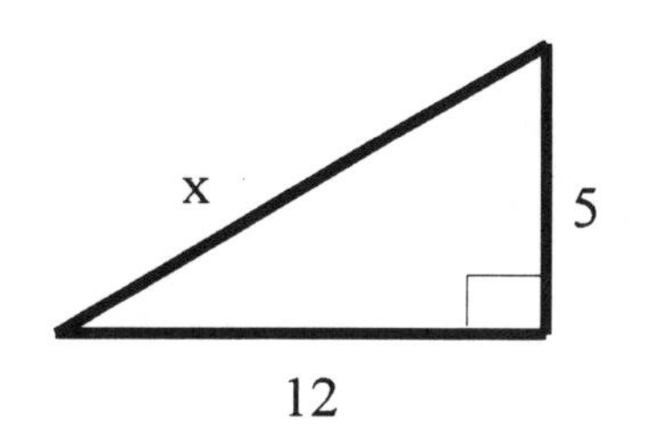

Step 1 The Pythagorean Triple 5, 12, 13 applies to this problem because: 1) two numbers from the Pythagorean Triple 5, 12, 13 are sides of the right triangle and 2) the greatest number, 13, in the Pythagorean Triple 5, 12, 13 will correspond to the hypotenuse.

Step 2 **x = 13** because 13 is the unused number from the Pythagorean Triple 5, 12, 13.

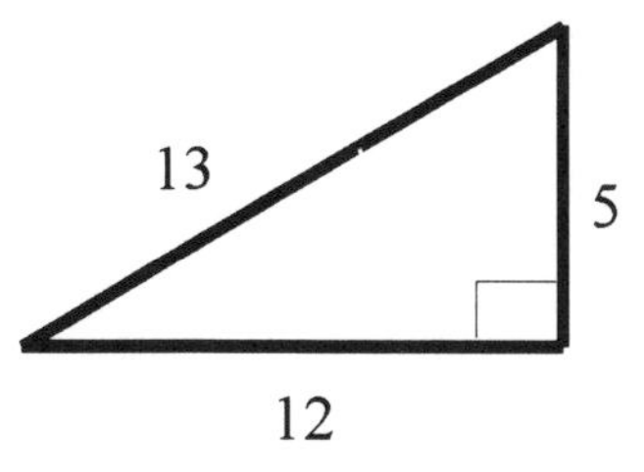

Example 3 Solve for x. Use a Pythagorean Triple if possible.

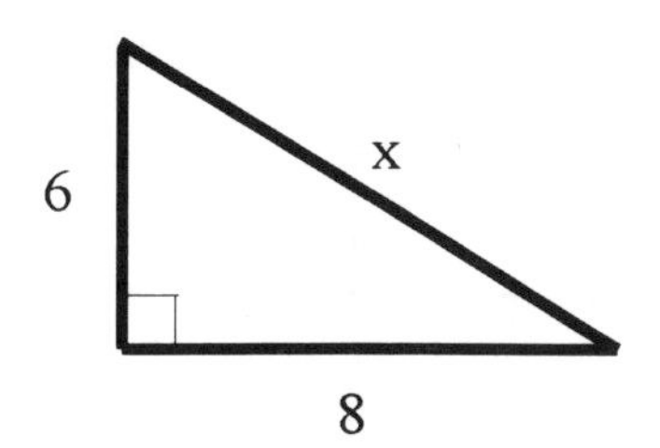

Step 1 The Pythagorean Triple 6, 8, 10 applies to this problem because: 1) two numbers of the Pythagorean Triple 6, 8, 10 are sides of the right triangle and 2) the greatest number, 10, in the Pythagorean Triple 6, 8, 10 will correspond to the hypotenuse.

Note: The Pythagorean Triple 6, 8, 10 was created by multiplying each number in the Pythagorean Triple 3, 4, 5 by 2.

Step 2 **x = 10** because 10 is the unused number from the Pythagorean Triple 6, 8, 10.

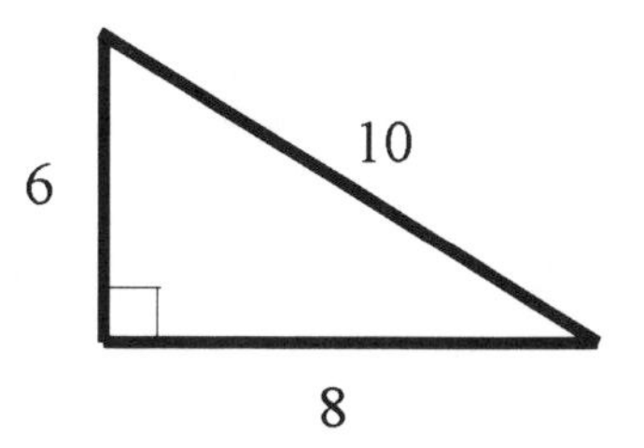

Example 4 **Solve for x. If possible use a Pythagorean Triple, otherwise use the Pythagorean Theorem.**

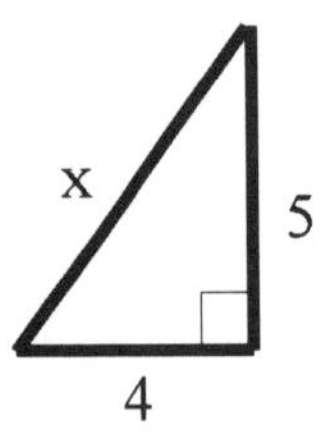

Step 1 The Pythagorean Triple 3, 4, 5 does **not** apply to this problem because the greatest number, 5, of the Pythagorean Triple 3, 4, 5 does **not** correspond to the hypotenuse.

Step 2 The **Pythagorean Theorem** applies to this problem because: 1) the triangle is a right triangle and 2) the length of two sides are known. In the Pythagorean Theorem ($a^2 + b^2 = c^2$) let $a = 4$ and $b = 5$ and solve for c. Recall that "c" **must always** represent the hypotenuse and "a" and "b" represent the legs of the right triangle in the formula $a^2 + b^2 = c^2$.

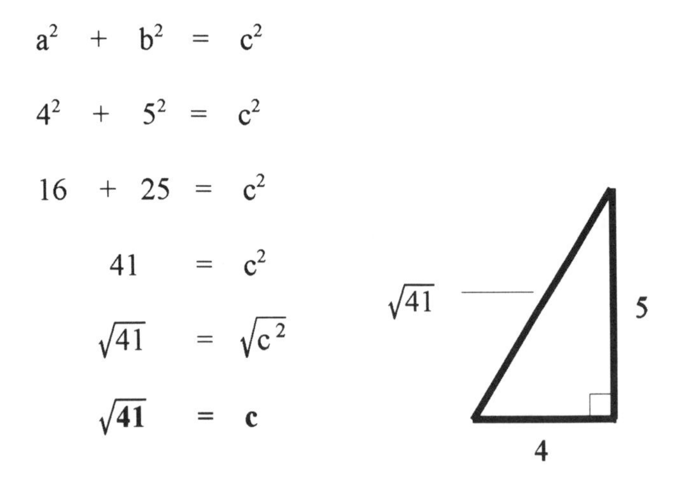

$$a^2 + b^2 = c^2$$

$$4^2 + 5^2 = c^2$$

$$16 + 25 = c^2$$

$$41 = c^2$$

$$\sqrt{41} = \sqrt{c^2}$$

$$\sqrt{41} = c$$

"x" is the hypotenuse, thus $\mathbf{x = \sqrt{41}}$ **.**

Note: A square root cancels out an exponent of 2, thus $\sqrt{c^2} = c$.

Exercises: **Solve for x by using Pythagorean Triples, if possible. If Pythagorean Triples do <u>not</u> apply, then use the Pythagorean Theorem.**

1)

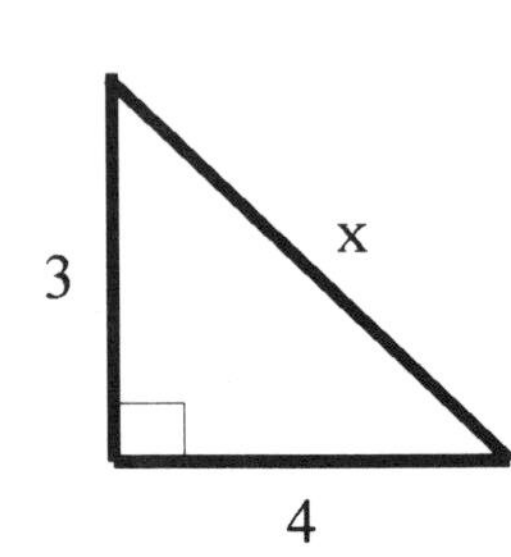

2)

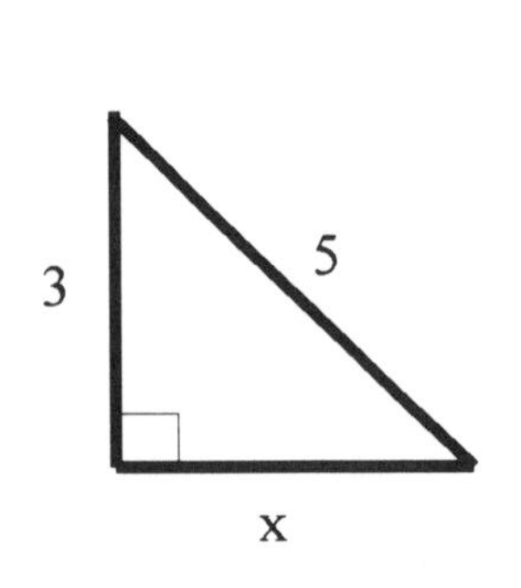

3)

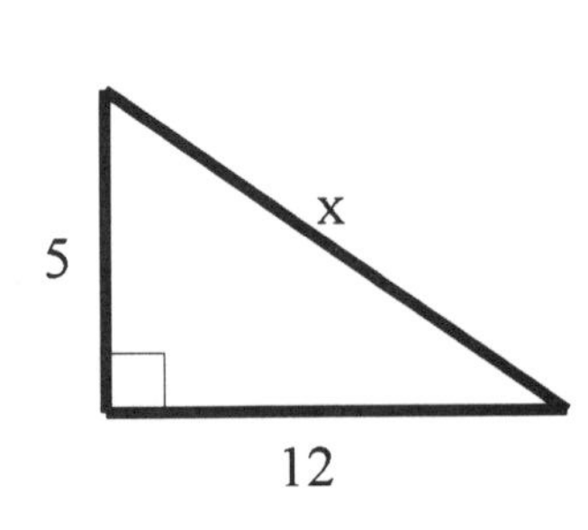

4)

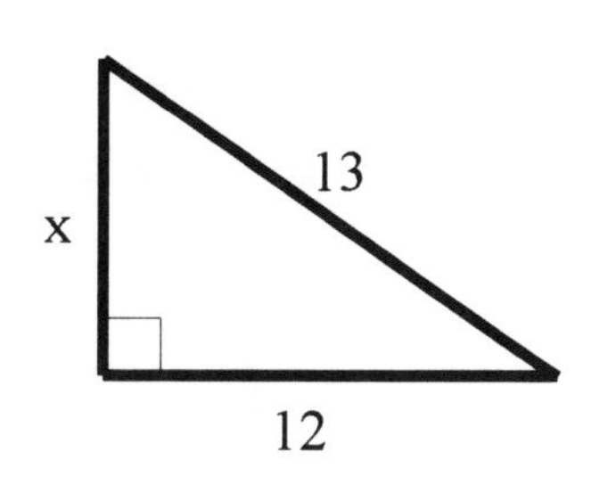

5)

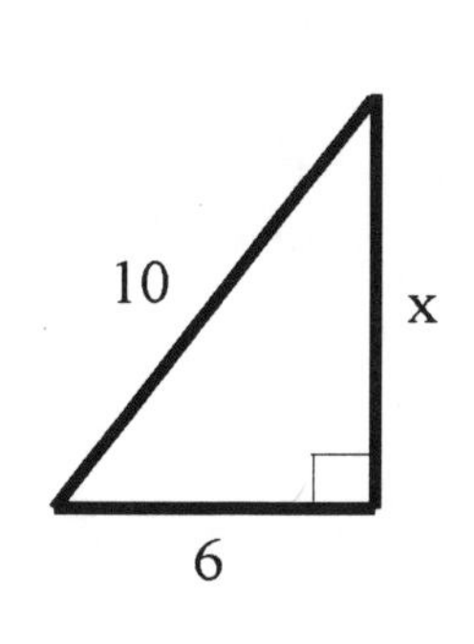

6)

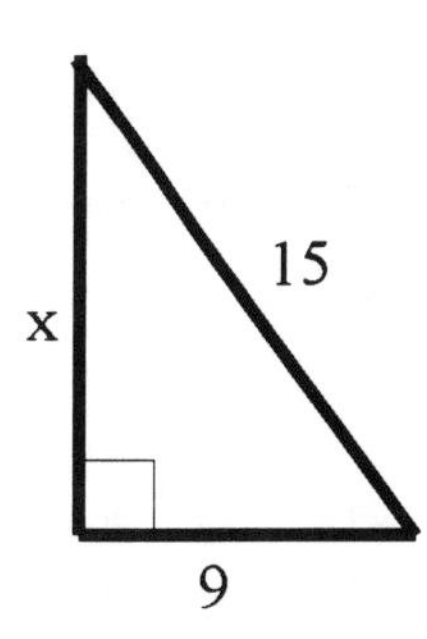

7)

10

8

x

8)

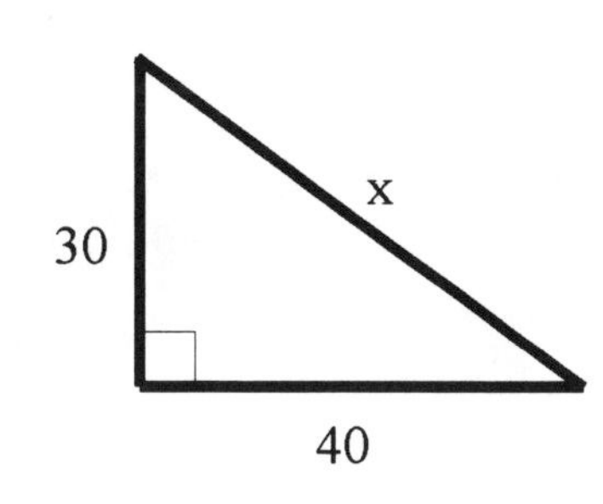

9)

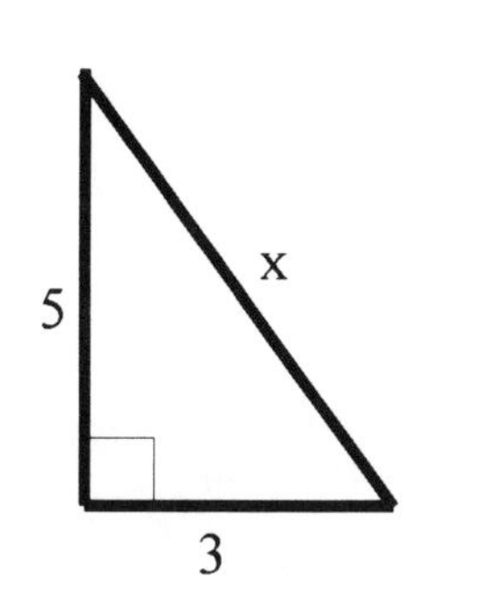

10)

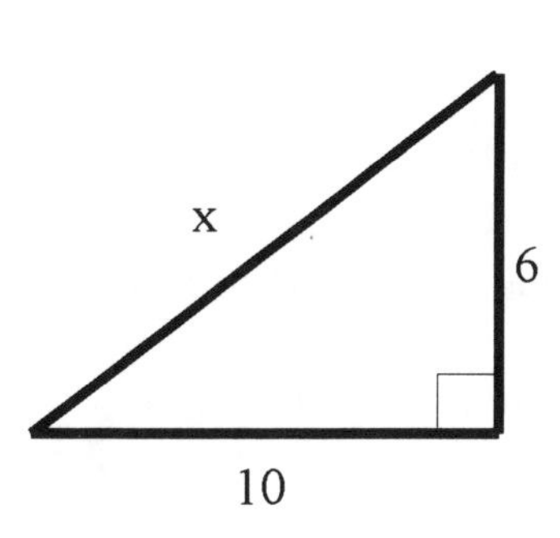

Shortcut #24 Determining a range of values

Substitute the minimum and maximum boundary values for the given variable (letter). This substitution is acceptable even if these boundary values are not included as values for the variable.

Example 1 **x and y form a straight line. The range of x is $40° < x < 60°$. Determine the range of y.**

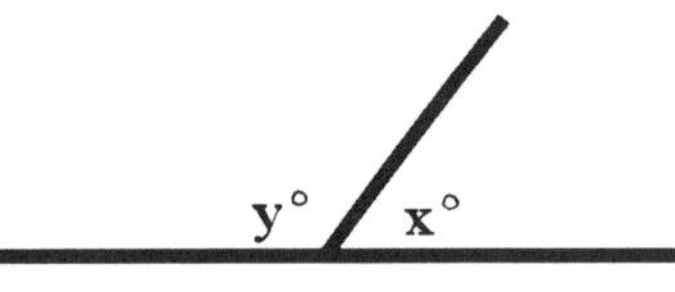

Step 1 Substitute the minimum boundary value for x, $40°$. If $x = 40°$, then $y = 140°$ because the sum of the angles around one side of a straight line is $180°$ ($40° + 140° = 180°$). Therefore, $140°$ is the upper limit of y ($y < 140°$). Notice that $140°$ is **not** a value of y because $40°$ is **not** a value of x.

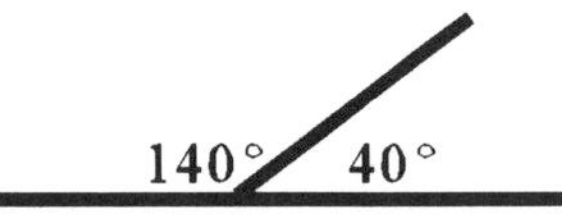

Note: Substitution of the **minimum** value for one variable will result in the **maximum** value of the other variable and conversely, substitution of the **maximum** value for one variable will result in the **minimum** value of the other variable.

Step 2 Substitute the maximum boundary value for x, $60°$. If $x = 60°$, then $y = 120°$ because the sum of the angles around one side of a straight line is $180°$ ($60° + 120° = 180°$). Therefore, $120°$ is the lower limit of y ($120° < y$). Notice that $120°$ is **not** a value of y because $60°$ is **not** a value of x.

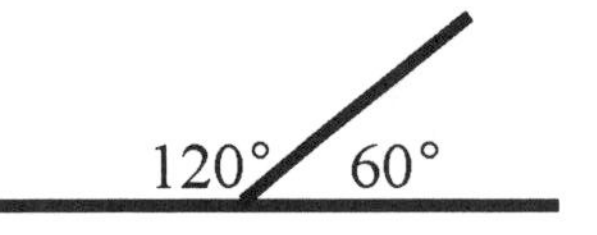

Step 3 Combine these two inequalities. This combined inequality is the range of possible values for y.

$120° < y$ and $y < 140°$ becomes $\mathbf{120° < y < 140°}$

Note: The larger value obtained will **always** be the upper boundary.

Example 2 **In triangle ABC, AB = AC. The range of x is $20° < x < 50°$.**
Determine the range of y.

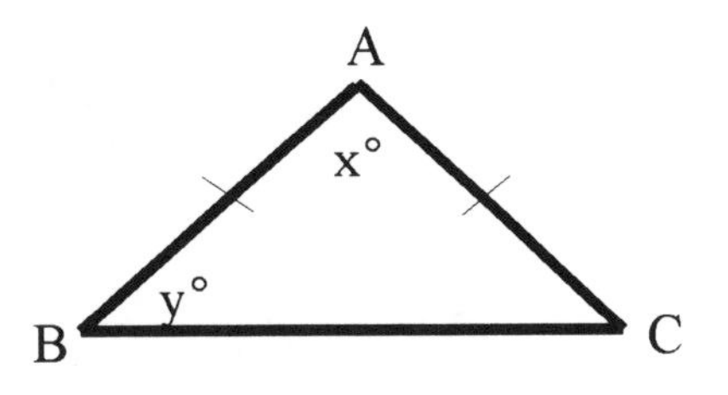

Step 1 Notice that $\angle B = \angle C$ because both angles are opposite equal sides. Substitute the minimum boundary value for x, 20°. If $x = 20°$, then $y = 80°$ because: 1) $\angle B = \angle C$ and 2) the sum of all three angles of any triangle is **always** 180° ($20° + 80° + 80° = 180°$). Therefore, 80° is the upper limit of y ($y < 80°$). Notice that 80° is **not** a value of y because 20° is **not** a value of x.

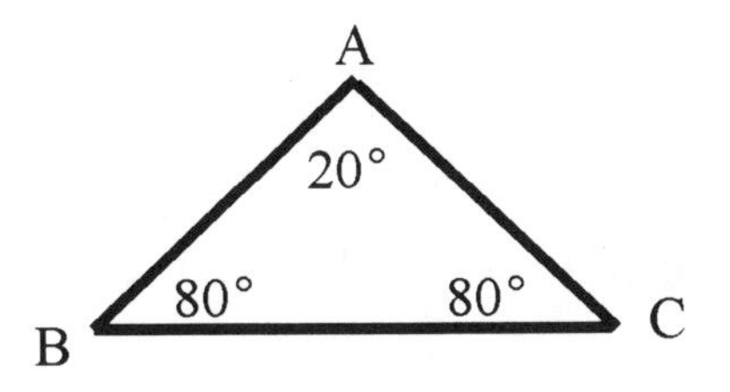

Step 2 Substitute the maximum boundary value for x, 50°. If $x = 50°$, then $y = 65°$ because: 1) $\angle B = \angle C$ and 2) the sum of all three angles of any triangle is **always** 180° ($50° + 65° + 65° = 180°$). Therefore, 65° is the lower limit of y ($65° < y$). Notice that 65° is **not** a value of y because 50° is **not** a value of x.

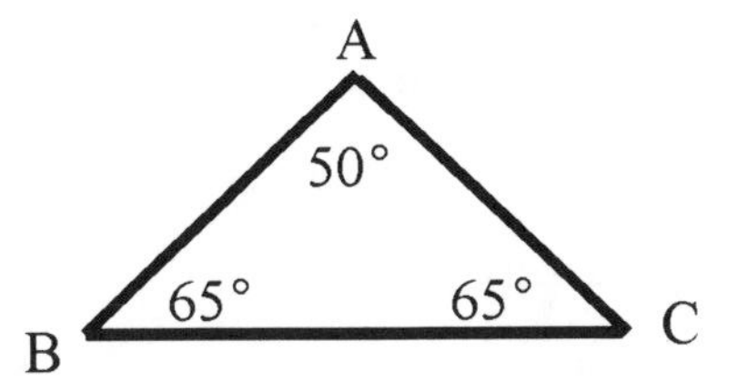

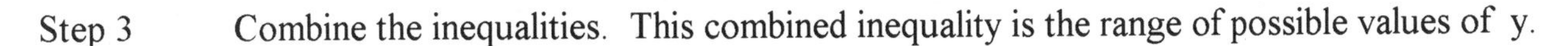

Step 3 Combine the inequalities. This combined inequality is the range of possible values of y.

$$65° < y \quad \text{and} \quad y < 80° \quad \text{becomes} \quad \mathbf{65° < y < 80°}$$

Exercises:

1) **Given: Range of x is $50° < x < 70°$**
Find: Range of y

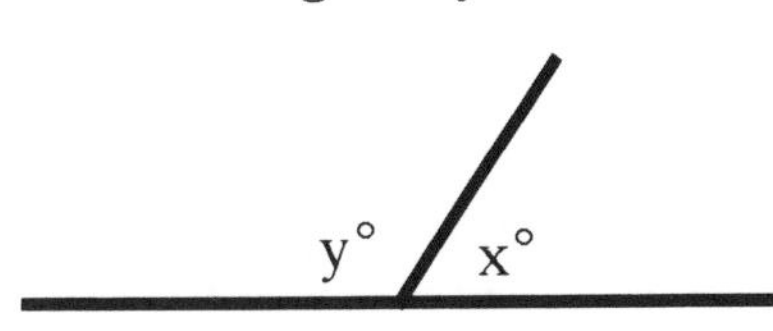

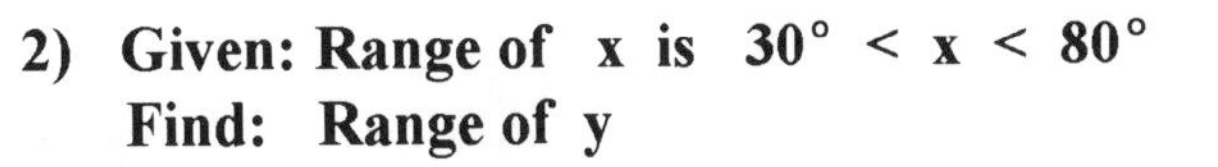

2) **Given: Range of x is $30° < x < 80°$**
Find: Range of y

y° x°

3) **Given: AB = AC, range of x is 40° < x < 70°**
Find: Range of y.

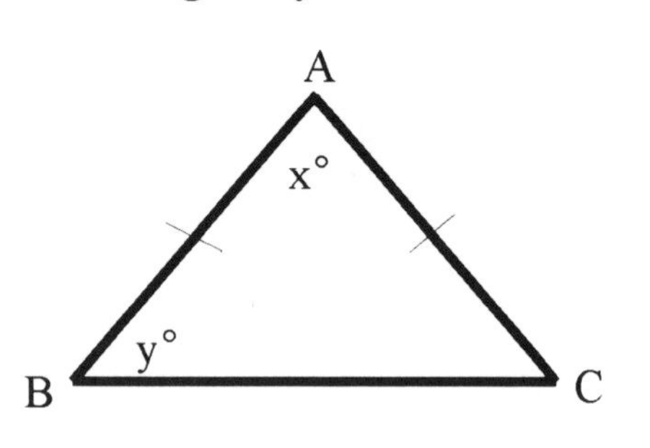

4) **Given: AB = AC, range of x is 20° < x < 60°. Find: Range of y.**

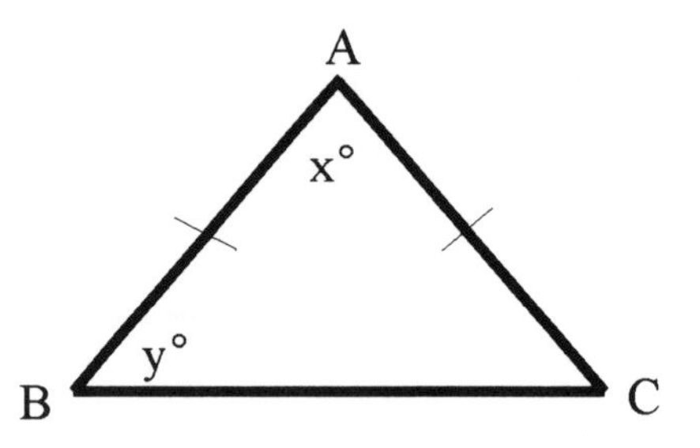

Word Problems: Shortcuts 25 - 34

Shortcut#25 Try the given answers

In some word problems, it is quicker to determine the correct answer by trying the given choices.

Example 1 **The sum of three consecutive numbers is 33. What is the middle number?**
Choices: 10, 11, 12, 13, 14

Step 1 Choose an answer that you think is correct and then test that possible answer. Here, 10 will be chosen first. If 10 is the middle number, the three consecutive numbers are 9, 10 and 11. If the sum of these three numbers is 33, then 10 is the correct answer.

$$9 + 10 + 11 = 30$$

The sum does **not** equal 33, so 10 is **not** the correct answer.

Step 2 Now choose 11. If 11 is the middle number, the three consecutive numbers are 10 , 11 and 12. If the sum of these three consecutive numbers is 33, then 11 is the correct answer.

$$10 + 11 + 12 = 33$$

The sum is 33, so **11** is the correct answer.

Example 2 **If Jim is twice as old as Jane and four years ago the sum of their ages was 52, how old is Jim now?**
Choices: 20, 30, 40, 50, 60

Step 1 Draw a chart representing Jim and Jane's present and past ages. Choose a likely answer for Jim's present age. 30 will be chosen first as Jim's present age. If Jim is now 30, then Jane is now 15. If the sum of their ages four years ago is 52, then Jim is now 30 years old.

	Present ages	Ages 4 years ago
Jim's age	30	26
Jane's age	15	11

Sum of ages 4 years ago: $26 + 11 = 37$

Jim's present age is **not** 30 because the sum of their ages four years ago is **not** 52.

Step 2 Jim's present age must be greater than 30 because the sum of 37 is less than the required sum of 52. Choose 40 for Jim's present age. If Jim is now 40, then Jane is now 20. If the sum of their ages four years ago is 52, then Jim's present age is 40.

	Present ages	Ages 4 years ago
Jim's age	40	36
Jane's age	20	16

Sum of ages 4 years ago: $36 + 16 = 52$

The sum of their ages four years ago is 52. Therefore, Jim's present age is **40**.

Exercises:

1) The sum of three consecutive integers is 15. What is the middle number?
Choices: 2, 3, 4, 5, 6

2) The sum of three consecutive integers is 93. What is the middle number?
Choices: 30, 31, 32, 33, 34

3) The sum of five consecutive numbers is 15. What is the greatest number?
Choices: 3, 4, 5, 6, 7

4) The ages of three brothers are consecutive even integers. The sum of their ages is 54. What is the age of the oldest brother now?
Choices: 16, 18, 20, 22, 24

5) If Jim is twice as old as Jane and two years ago the sum of their ages was 41, how old is Jim now?
Choices: 15, 20, 25, 30, 35

6) A father is three times as old as his son. Five years ago the sum of their ages was 30. How old is the father now?
Choices: 30, 35, 40, 45, 50

7) A mother is four times as old as her daughter. In two years, the sum of their ages will be 44. How old is the mother now?
Choices: 30, 32, 34, 36, 38

Shortcut #26 "OF" means multiply

The word "of" indicates multiplication, if it is preceded by a fraction, percent or decimal numeral.

Example 1 $\frac{2}{9}$ *of* $\frac{4}{7}$ =

Step 1 Substitute a multiplication sign for the word "of."

$$\frac{2}{9} \times \frac{4}{7} =$$

Step 2 Multiply the fractions.

$$\frac{2}{9} \times \frac{4}{7} = \frac{8}{63}$$

Example 2 $\mathbf{\frac{1}{2}}$ ***of*** $\mathbf{\frac{3}{5}}$ ***of*** **3** =

Step 1 Substitute a multiplication sign for the word "of ."

$$\frac{1}{2} \times \frac{3}{5} \times 3 =$$

Step 2 Write 3 as $\frac{3}{1}$ and then multiply.

$$\frac{1}{2} \times \frac{3}{5} \times \frac{3}{1} = \mathbf{\frac{9}{10}}$$

Example 3 **15% of 20 =**

Step 1 Substitute a multiplication sign for the word "of."

$$15\% \times 20 =$$

Step 2 Before a percent is multiplied, it must be converted into a fraction or decimal numeral. To convert a percent into a fraction or decimal numeral, divide the percent by 100. Dividing a decimal numeral by 100 may be accomplished by moving the decimal point two places to the left. If a decimal point does **not** exist, place a decimal point at the far right of the number.

$$15\% = 15.0\% = .15 \quad \text{OR} \quad 15\% = \frac{15}{100}$$

Step 3 Substitute .15 for 15% and then multiply.

$$.15 \times 20 = \mathbf{3}$$

Example 4 **30% of 40% =**

Step 1 Substitute a multiplication sign for the word "of."

$$30\% \times 40\% =$$

Step 2 Convert all percents to decimal numerals or fractions and then multiply. Recall from example 3, that dividing a number by 100 is equivalent to moving its' decimal point two places to the left. Also note that .30 = .3 and .40 = .4 because zeros at the far right of a decimal numeral may be eliminated.

$$30\% = 30.0\% = .30 = .3 \qquad 40\% = 40.0\% = .40 = .4$$

$$.3 \times .4 = \mathbf{.12}$$

Note: If percents are the only quantities being multiplied, the answer may be written as percent. Any decimal numeral or fraction may be converted into a percent by multiplying it by 100. Multiplying a number by 100 is equivalent to moving its' decimal point two places to the right. Thus, .12 = 12.0% = 12%

Exercises: **Solve and reduce fractions when possible.**

1) $\frac{2}{5}$ *of* 30 = **2)** $\frac{3}{7}$ *of* 42 = **3)** $\frac{5}{6}$ *of* 7 = **4)** $\frac{3}{8}$ *of* $\frac{4}{9}$

5) $\frac{1}{6}$ *of* $2\frac{1}{3}$ = **6)** $\frac{3}{4}$ *of* $\frac{1}{2}$ *of* 3 = **7)** $\frac{2}{5}$ *of* $\frac{1}{3}$ *of* $\frac{3}{8}$

8) $\frac{1}{2}$ *of* $\frac{1}{2}$ *of* 12 = **9)** 20% *of* 75 = **10)** 35% *of* 10 =

11) 6.9% *of* $15 =$ **12)** 20% *of* $25\% =$ **13)** 35% *of* $60\% =$

14) 10% *of* 10% *of* $50 =$

Shortcut #27 Substitute 100 for an unknown quantity value that is changed by a percent

If an unknown quantity value is increased or decreased by a percent, substitute 100 for the quantity value and then calculate the new value.

Note: In quantitative comparisons, numbers other than 100 may also need to be substituted. This is always the case if more than 1 unknown quantity is changed by a percent.

Example 1 **Car production in 1996 is 20% greater than in 1990. Truck production in 1996 is 30% less than in 1990. The ratio of car to truck production in 1996 is how many times the ratio of car to truck production in 1990?**

Step 1 Car and truck production in 1990 are the unknown quantities that are changed by a percent. Therefore, assume car and truck production to both be 100 in 1990.

Step 2 Calculate car and truck production in 1996. Car production is increased by 20%, thus multiply car production in 1990 by 120% (100% + 20% = 120%). Truck production is decreased by 30%, thus multiply truck production in 1990 by 70% (100% - 30% = 70%).

1996 Car Production:

$100 \times 120\% =$
$100 \times 1.2 = 120$

1996 Truck Production:

$100 \times 70\% =$
$100 \times .7 = 70$

Step 3 Calculate the ratio of car to truck production in 1990 and in 1996.

Ratio of Car to Truck Production: 1990: 1996:

$$\frac{\textit{Car Production}}{\textit{Truck Production}} = \qquad \frac{100}{100} = 1 \qquad \frac{120}{70} = \frac{12}{7}$$

Step 4 Calculate how many times the 1996 ratio is compared to the 1990 ratio. In other words, what number must the 1990 ratio be multiplied by to obtain the 1996 ratio.

$$\text{1990 ratio} \times \underline{\ ?\ } = \text{1996 ratio}$$

$$1 \times \underline{\ ?\ } = \frac{12}{7}$$

"1" must be multiplied by $\frac{12}{7}$ to obtain $\frac{12}{7}$, thus the 1996 ratio is $\frac{12}{7}$ times the 1990 ratio.

Example 2 The length and width of rectangle A are 20% less and 40% less, respectively, then the length and width of rectangle B. The perimeter of rectangle A is what percent of the perimeter of rectangle B?

Step 1 The length and width of rectangle B are the unknown quantities that are changed by a percent. Therefore, assume the length and width of rectangle B to both equal 100. Notice that if the length equals the width, the figure is a square. This is acceptable because a square is a rectangle.

Rectangle B:

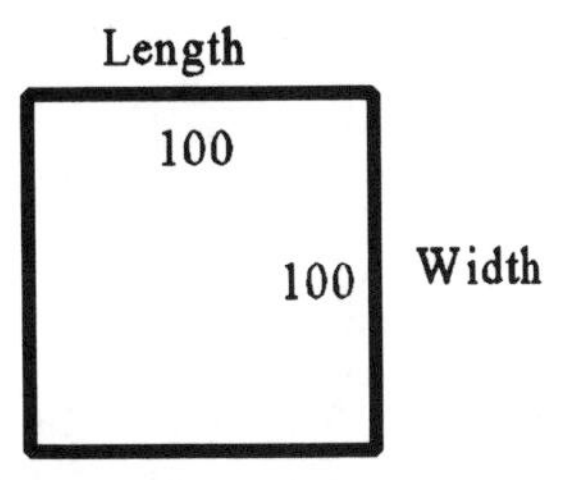

Step 2 Calculate the length and width of rectangle A. The length of rectangle A is 20% less then the length of rectangle B. Thus, multiply the length of rectangle B by 80% (100% - 20% = 80%). The width of rectangle A is 40% less than the width of rectangle B. Thus, multiply the width of rectangle B by 60% (100% - 40% = 60%).

Length of Rectangle A:

100 x 80% =
100 x .8 = 80

Width of Rectangle A:

100 x 60% =
100 x .6 = 60

Rectangle A:

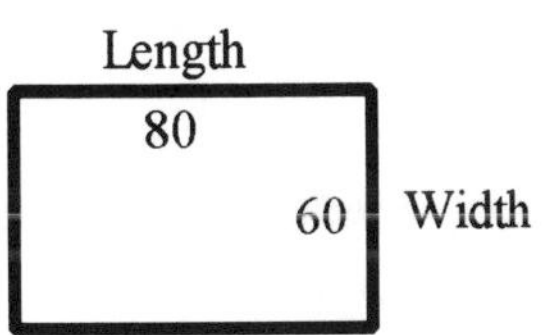

Step 3 Determine the perimeter of rectangle A and rectangle B.

Perimeter of Rectangle A:
100 + 100 + 100 + 100 = 400

Perimeter of Rectangle B:
80 + 80 + 60 + 60 = 280

Step 4 Determine what percent the perimeter of rectangle A to be of the perimeter of rectangle B. In other words, the perimeter of rectangle A is what part of the perimeter of rectangle B. This can be represented as a fraction and then converted into a percent.

$$\frac{\textit{Perimeter of Rectangle A}}{\textit{Perimeter of Rectangle B}} = \frac{280}{400} = \frac{28}{40} = .7$$

To convert a decimal or fraction into a percent, multiply by 100.
(.7) x 100 = 70%, thus the perimeter of rectangle A is 70% of the perimeter of rectangle B.

Exercises:

1) Car production in 1995 was 40% greater than in 1990. Truck production in 1995 was 40% less than in 1990. The ratio of car to truck production in 1995 is how many times the ratio of car to truck production in 1990?

2) Car production in 1995 was 200% greater than in 1970. Truck production in 1995 was 50% less than in 1970. The ratio of car to truck production in 1995 is how many times the ratio of car to truck production in 1970?

3) Car production in 1945 was 50% less than in 1940. Truck production in 1945 was 50% greater than in 1940. The ratio of car to truck production in 1945 is how many times the ratio of car to truck production in 1940?

4) The length and width of rectangle A are 10% less and 20% less, respectively, then the length and width of rectangle B. The perimeter of rectangle A is what percent of the perimeter of rectangle B?

5) The length and width of rectangle A are 30% greater and 20% greater, respectively, then the length and width of rectangle B. The perimeter of rectangle A is what percent of the perimeter of rectangle B?

6) The length and width of rectangle A are each 50% less then the length and width of rectangle B. The area of rectangle A is what percent of the area of rectangle B?

Shortcut #28 Determining how many consecutive numbers are in a group

Subtract the largest number by the smallest number and then add 1.

Note: **Consecutive numbers** are numbers ordered one after the other without interruption. Examples of consecutive numbers are: (1, 2, 3, 4) or (7, 8, 9) or (7, 8, 9...). Notice that consecutive numbers may or may not continue indefinitely.

Example **Customers numbered 3 through 76 are waiting in line. How many customers are waiting in line?**

Step 1 Subtract the largest number by the smallest number and then add 1.

$$76 - 3 = 73 \rightarrow 73 + 1 = \mathbf{74}$$

Thus, there are **74 customers** waiting in line.

Exercises:

1) Customers numbered 3 through 7 are waiting in line. How many customers are waiting in line?

2) Customers numbered 1 through 10 are waiting in line. How many customers are waiting in line?

3) Customers numbered 5 through 61 are waiting in line. How many customers are waiting in line?

4) Customers numbered 23 through 112 are waiting in line. How many customers are waiting in line?

5) In an automobile factory, cars numbered 56 through 1020 must be shipped today. How many cars must be shipped today?

6) How many numbers are in the list 312 through 684, inclusive*?

*Note: **"Inclusive"** means the first and last numbers are included in the group.

Shortcut #29 Arrangements

An arrangement is a group of items. A new arrangement is created if one or more items in an arrangement are changed or if the order of the same items in an arrangement is changed. The total number of arrangements is determined by multiplying a series of numbers together.

Example 1 **Four people are waiting in line. How many arrangements of their order exist?**

Step 1 Notice that the question asks for arrangements. Recall that a different order creates a new arrangement. The number of possible arrangements is found by multiplying a series of numbers. Draw four blanks. Each blank represents the four positions in the line: first, second, third and fourth.

____	____	____	____
First	Second	Third	Fourth

Step 2 Fill in the first blank with a 4. This represents the fact that any one of the four people could be first in line.

4	____	____	____
First	Second	Third	Fourth

Step 3 Now that the first position in line is filled, three people remain to be placed in line. Any one of these three people could be second in line, so write a 3 in the second position.

4	3	____	____
First	Second	Third	Fourth

Step 4 The first and second positions in the line are now filled. Two people remain to be placed in the line. Because any one of these two people could be third in line, place a 2 in the third position.

4	3	2	____
First	Second	Third	Fourth

Step 5 Only one person remains to be placed in line. The only position available is the fourth position, so place a 1 in the fourth position.

4	3	2	1
First	Second	Third	Fourth

Step 6 Multiply these four numbers together to determine the total number of possible arrangements.

$4 \times 3 \times 2 \times 1 =$ **24 arrangements**

Example 2 **How many two letter arrangements exist from the letters: A, B, C, D, and E? No letter may be used more than once in each arrangement.**

Step 1 Notice that the question asks for arrangements. Recall that different items **or** a different order of the same items creates a new arrangement. The number of possible arrangements is found by multiplying a series of numbers. Draw two blanks, one for each of the two positions in the arrangement.

____	____
First	Second

Note: Notice that the number of blanks drawn does **not** have to equal the number of given items.

Step 2 Fill in the first blank with a 5, because any one of the five letters could be written first.

5	____
First	Second

Step 3 Now that the first position is filled, there remain four letters to be positioned second. Any one of these four letters could be placed in the second position, therefore place a 4 in the second position.

Note: If any letter could be used twice, a 5 would be placed in the second position because any of the 5 letters could be in the second position.

5	4
First	Second

Step 4 Multiply these two numbers together to determine the total number of arrangements.

5 x 4 = **20 arrangements**

Exercises:

1) Three people are waiting in line. How many different ways can their order be arranged?

2) Five people are waiting in line. How many different ways can their order be arranged?

3) Six people are waiting in line. How many different ways can their order be arranged?

In exercises 4 – 9, no letter may be used more than once in each arrangement.

4) How many three letter arrangements exist from the letters: A, B, and C?

5) How many four letter arrangements exist from the letters: A, B, C, and D?

6) How many two letter arrangements exist from the letters: A, B, C, and D?

7) How many three letter arrangements exist from the letters: A, B, C, and D?

8) How many two letter arrangements exist from the letters: A, B, C, D, E, and F?

9) How many four letter arrangements exist from the letters: A, B, C, D, E, and F?

Shortcut #30 Combinations from one group

A combination is a group of items. A new combination is created only if one or more items in a combination are changed. A different order of the same items does not create a new combination. The total number of combinations that contain only two items is determined by adding a series of numbers.

Note: **This procedure applies only if the combinations to be created contain only two items.**

Example 1 **Five people will shake each others hand. How many handshakes will occur?**

Step 1 Notice that combinations are to found because a different order of the same two people shaking hands does **not** create a new handshake. Recall that the number of combinations containing two items is found by **adding** a series of numbers. In this example, the two items being combined are the two people shaking hands.

Step 2 List the five people in order.

Person:	1st	2nd	3rd	4th	5th

Step 3 List the number times each person shakes hands and do this in order. The first person shakes hands with the other four people, so place a 4 underneath the first person.

Person:	1st	2nd	3rd	4th	5th
Handshakes:	4				

Step 4 Now count the remaining handshakes the second person will make. The second person will shake the hand of the last three people. Therefore, place a 3 underneath the second person. Notice that the second person has already shaken the first person's hand, so this is why that handshake is **not** counted in this step.

Person:	1st	2nd	3rd	4th	5th
Handshakes:	4	3			

Step 5 The third person needs only to shake hands with the last two people. The fourth person will then need to shake only the last person's hand and the last person will then have shaken everyone's hand. Place these number of handshakes underneath the appropriate person.

Person:	1st	2nd	3rd	4th	5th
Handshakes:	4	3	2	1	0

Step 6 **Add** all these handshakes to determine the total number of handshakes made by these five people.

$4 + 3 + 2 + 1 + 0 =$ **10 handshakes**

Example 2 **From seven different colors, any two can be mixed together. How many mixtures can be created?**

Step 1 Notice that combinations are to be found because a different order of the same two colors mixed together does **not** create a new mixture. Recall that the number of combinations containing two items is found by **adding** a series of numbers. In this example, the two items being combined are two colors to create a new mixture. Follow the same procedure used in example 1. First list the seven given colors by number.

Colors:	1st	2nd	3rd	4th	5th	6th	7th

Step 2 The first color can be mixed with the six other colors, so place a 6 underneath the first color.

Colors:	1st	2nd	3rd	4th	5th	6th	7th
Pairs:	6						

Step 3 Because the second color has already been mixed with the first color, the second color need only be mixed with the last five colors. Therefore, place a 5 below the second color.

Colors:	1st	2nd	3rd	4th	5th	6th	7th
Pairs:	6	5					

Step 4 The third color will only need to mixed with the last 4 colors. The fourth color will only to be mixed with the last 3 colors. The fifth color need only be mixed with the last 2 colors. The sixth color need only be mixed with the last color. The seventh color has now been mixed with all the other colors. Place the number of pairs below the appropriate color.

Colors:	1st	2nd	3rd	4th	5th	6th	7th
Pairs:	6	5	4	3	2	1	0

Step 5 **Add** the number of pairs to obtain the total possible number of pairs.

$$6 + 5 + 4 + 3 + 2 + 1 + 0 = \mathbf{21} \textbf{ pairs}$$

Exercises:

1) Four people will shake each others hand. How many handshakes will occur?

2) Eight people will shake each others hand. How many handshakes will occur?

3) From five different colors, any two can be mixed together. How many mixtures can be created?

4) From three different colors, any two can be mixed together. How many mixtures can be created?

5) From six different colors, any two can be mixed together. How many mixtures can be created?

Shortcut #31 Combinations from two or more groups

If one item is chosen from each group, then the total number of combinations is the product of the number of items in each group.

Note: A <u>combination</u> is a group of items. A new combination is created only if one or more items in the combination are changed. A different order of the <u>same</u> items does <u>not</u> create a new combination.

Example 1 **A person has 5 shirts and 3 pairs of pants. How many different outfits can be created using 1 shirt and 1 pair of pants?**

Step 1 Notice that this is a combination problem because a different order of the same items does **not** create a new outfit. If one item is chosen from each group, the total number of combinations is the product of the number of items in each group. One item is to be chosen from each of the two groups: shirts and pants.

5 shirts x 3 pants = **15 combinations (outfits)**

Example 2 **A person has 3 hats, 4 jackets and 6 ties. How many different outfits can be created using 1 hat, 1 jacket and 1 tie?**

Step 1 Notice that this is a combination problem because a different order of the same items does **not** create a new outfit. If one item is chosen from each group, the total number of combinations is the product of the number of items in each group. One item is to be chosen from each of the three groups: hats, jackets and ties.

3 hats x 4 jackets x 6 ties = **72 combinations (outfits)**

Exercises:

1) A person has 12 shirts and 6 pair of pants. How many different outfits can be created using 1 shirt and 1 pair of pants?

2) A person has 5 shirts, 3 jackets and 4 ties. How many different outfits can be created using 1 shirt, 1 jacket and 1 tie?

Shortcut #32 Convert quantities by cancellation of labels

Multiply the quantity to be converted by one or more fractions whose value equals 1, in order to cancel unwanted labels and to input the desired label.

Notes on cancellation of labels:

1) Fractions whose numerator and denominator are equivalent are equal to 1. Thus, the fractions: (60 mins.) / (1 hr.) or (1 hr.) / (60 mins.) each have a value of 1 because 60 mins. = 1 hr. Any quantity multiplied by one or more such fractions will **not** change in value. However, the answer will have a different label and a different number. Therefore, this procedure may be used to convert the label (also known as units) of any quantity.

2) Identical labels may be canceled **only** if one label is in a numerator and the other identical label is in a denominator.

3) If identical labels are canceled, the numbers preceding those labels do **not** have to be canceled.

4) Labels that are **not** canceled remain as labels of the answer. If the answer requires only one label, then that label **must** be in the numerator of one of the fractions being multiplied together.

Note: Conversions can also be accomplished by using proportions. However, cancellation of labels is a superior technique when quantities must be converted more than once as required in examples 3 and 4. In example 1 is solved using each method.

Example 1 3 hours equals how many minutes?

Step 1 Multiply 3 hrs. by the fraction (60mins.) / (1 hr.). Multiplying by the fraction (60 mins.) / (1 hr.) will **not** change the value of 3 hours because 60 mins. = 1 hour (60 mins. / 1 hr.= 1). The fraction (60 mins.) / (1 hr.) is used rather than its' reciprocal (1 hr.) / (60 mins.) because: 1) in order to cancel the label of hours, one label of hours **must** be in a numerator and the other label of hours **must** be in a denominator and 2) the answer requires a label of minutes, therefore the label of minutes **must** be in a numerator of one of the fractions being multiplied.

$$(3 \text{ hrs.}) \frac{(60 \text{ mins.})}{(1 \text{ hr.})} =$$

Note: A denominator of 1 may be placed below any quantity without changing its' value. Therefore, (3 hrs.) may be written as (3 hrs.) / 1 in order to clearly see that (3 hrs.) is in the numerator.

Step 2 Cancel the label of hours. Notice that the numbers preceding these labels do **not** have to be canceled. Multiply all the numbers. The label of the answer is minutes because this label is **not** canceled and because this label is in a numerator of a fraction being multiplied. Thus, 3 hours equals 180 minutes.

$$\frac{(3 \cancel{\text{hrs.}})}{(1)} \frac{(60 \text{ mins.})}{(1 \cancel{\text{hr.}})} = \mathbf{180 \text{ mins.}}$$

Alternative Solution using a Proportion:

Step 1 Set up a proportion and label the unknown quantity as x. The labels in the numerators must be identical and the labels in the denominators must be identical. Also notice that the numerator and denominator of each fraction must be equivalent, for example, 3 hours will equals x minutes and 1 hour equals 60 minutes.

Note: A proportion can be created in a number of other ways because the numerator of either fraction can be interchanged with the denominator of the other fraction.

$$\frac{3\ hrs.}{x\ mins.} = \frac{1\ hr.}{60\ mins.}$$

Step 2 Cross-multiply and solve the equation for x.

$$(3)(60) = (1)(x)$$

$$180 = x$$

Thus, 3 hrs. equals 180 mins.

Example 2 70 minutes equals how many hours?

Step 1 Multiply 70 mins. by the fraction (1hr.) / (60 mins.). 70 minutes may be written with a denominator of 1.

$$\frac{(70\ \text{mins.})}{(1)} \ \frac{(1\ \text{hr.})}{(60\ \text{mins.})} =$$

Step 2 Cancel the label of minutes. Multiply and divide the numbers. The answer will have the label of hours because this label is **not** canceled and because the label of hours is in a numerator of one of the fractions being multiplied. Thus, 70 minutes equals 1 1/6 hours.

$$\frac{(70\ \not{\text{mins}}.)}{(1)} \ \frac{(1\ \text{hr.})}{(60\ \not{\text{mins}}.)} = \frac{70}{60} = \frac{7}{6} = 1\frac{1}{6}\ \text{hours}$$

Example 3 2 hours equals how many seconds?

Step 1 In this conversion, 2 hours must be multiplied by two fractions. First convert hours to minutes by multiplying 2 hours by the fraction (60 mins.) / (1 hour). 2 hours may be written with a denominator of 1.

$$\frac{(2\ \text{hrs.})}{(1)} \ \frac{(60\ \text{mins.})}{(1\ \text{hr.})} =$$

Step 2 Now multiply by the fraction (60 secs.) / (1 min.) in order to convert minutes to seconds.

$$\frac{(2 \text{ hrs.})}{(1)} \; \frac{(60 \text{ mins.})}{(1 \text{ hr.})} \; \frac{(60 \text{ secs.})}{(1 \text{ min.})} =$$

Step 3 Cancel the labels of hours and minutes. Multiply the numbers. Notice that the answer will have the label of seconds because this label is **not** canceled and this label is in a numerator of one of the fractions being multiplied. Thus, 2 hours equals 7,200 seconds.

$$\frac{(2 \not{\text{hrs}}.)}{(1)} \; \frac{(60 \not{\text{mins}}.)}{(1 \not{\text{hr}}.)} \; \frac{(60 \text{ secs.})}{(1 \not{\text{min}}.)} = \textbf{7,200 seconds}$$

Example 4 Two books cost "c" cents. How many books can be bought for "d" dollars?

Step 1 This conversion requires multiplication by two additional fractions. Write the fraction (2 books) / (c cents). The quantity of books is placed in the numerator because the question asks for a number of books. Multiply this fraction by the fraction (100 cents) / (1 dollar). Notice that 100 cents equals 1 dollar and that 100 cents is written in the numerator in order to cancel the label of cents. The result is the cost of two books in dollars.

$$\frac{(2 \text{ books})}{(c \text{ cents})} \; \frac{(100 \text{ cents})}{(1 \text{ dollar})} =$$

Step 2 Now multiply by "d" dollars. Notice that the label of dollars will cancel out because one label is in a numerator and the other is in a denominator. "d" dollars may be written with a denominator of 1.

$$\frac{(2 \text{ books})}{(c \text{ cents})} \; \frac{(100 \text{ cents})}{(1 \text{ dollar})} \; \frac{(d \text{ dollars})}{(1)} =$$

Step 3 Cancel the labels of cents and dollars. Notice that the remaining label is "books" and this label is in a numerator, therefore "books" will be the label of the answer. Multiply the numbers and variables (letters) together. (200d) / (c) books can therefore be bought for "d" dollars.

$$\frac{(2 \text{ books})}{(c \not{\text{cents}})} \; \frac{(100 \not{\text{cents}})}{(1 \not{\text{dollar}})} \; \frac{(d \not{\text{dollars}})}{(1)} = \frac{200d}{c} \text{ books}$$

Exercises:

1) 7 hours equals how many minutes?

2) 135 minutes equals how many hours?

3) 3 hours equals how many seconds?

4) 360 seconds equals how many hours?

5) 15 books cost "d" dollars. How many books can be bought for 12 dollars?

6) 10 books cost "c" cents. How many books can be bought for "d" dollars?

7) 5 books cost "c" cents. How many books can be bought for "d" dollars?

Shortcut #33 Intersections of two categories

An <u>intersection</u> of two categories is the quantity in <u>both</u> categories.
To solve this type of problem perform the following two steps:

1st) Determine the number in <u>both</u> categories.

2nd) Determine the number in <u>only</u> a single category (do this for each category).

Example 1 **There are 9 students in the Math class. There are 5 students in the French class. If 3 students take <u>both</u> math and French, what is the total number of students taking one or both classes?**

Step 1 Draw two overlapping circles. One circle represents the number of students taking math and the other circle represents the number of students taking French. The overlapping section represents the number of students taking **both** math and French. The non-overlapping sections represent the number of students taking only one of these classes.

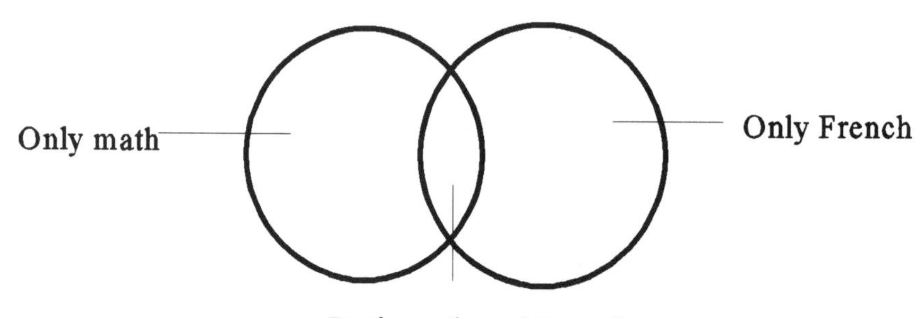

Step 2 In the overlapping section, fill in the number of students taking **both** math and French (3 students).

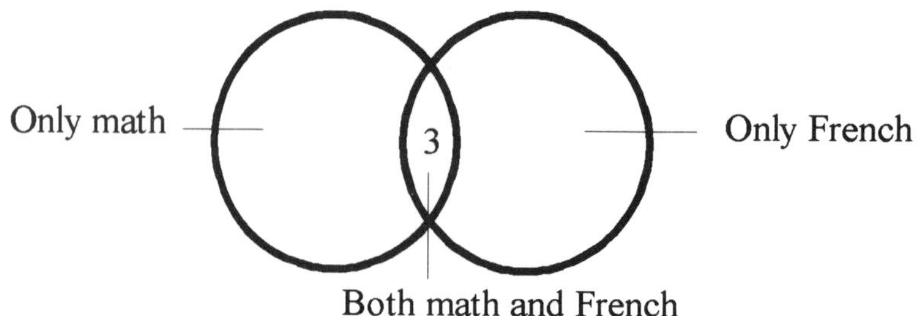

Step 3 In the non-overlapping sections, fill in the number of students taking **only** math or **only** French. Remember that the total number of students in the math class **must** be 9 and the total number of students in the French class **must** be 5. The number of students taking **only** math is 6 (9 – 3 = 6). The number of students taking **only** French is 2 (5 – 3 = 2).

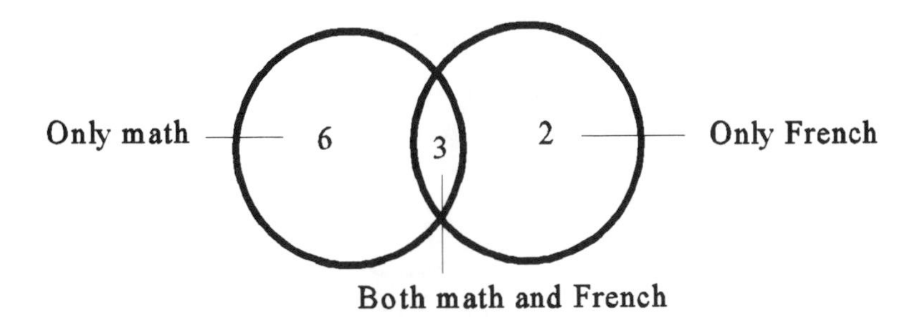

The picture now shows that 9 students are in the math class (6 + 3 = 9) and 5 students are in the French class (2 + 3 = 5). The picture also shows that 6 students take **only** math, 2 students take **only** French and 3 students take **both** math and French.

Step 4 The total number of students in one or both classes is the sum of the three numbers in this picture. The total number of students in one or both classes is: 6 + 3 + 2 = **11** .

Example 2 Of 13 students, each student takes French or Spanish or both classes. 10 students are in the French class and 7 students are in the Spanish class. How many students are in both classes? Choices: 1, 2, 3, 4, 5

Step 1 Draw two overlapping circles. One circle represents the number of students in the French class and the other circle represents the number of students in the Spanish class.
The overlapping section represents the number of students in **both** classes.
The non-overlapping sections represent the number of students taking **only** one class or the other.

Step 2 Choose a likely answer for the number of students in both classes and write this number in the overlapping section. First, 2 will be chosen.

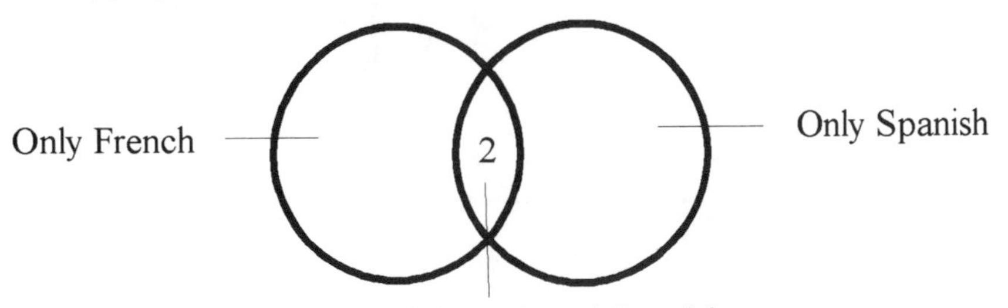

Step 3 Fill in the remaining number of students taking **only** one class or the other. Remember that the French class **must** have a total of 10 students and the Spanish class **must** have a total of 7 students. If two students are in both classes, then the number of students taking **only** French is 8 (10 – 2 = 8) and the number of students taking **only** Spanish is 5 (7 – 2 = 5).

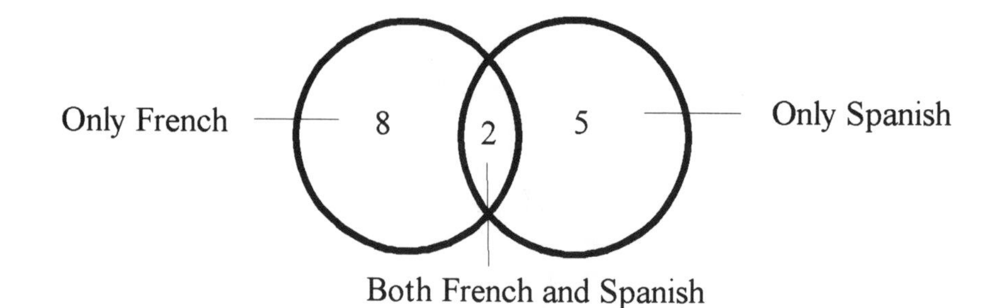

The picture now shows that the French class has 10 students (8 + 2 = 10) and the Spanish class has 7 students (5 + 2 = 7). The picture also shows that 8 students take **only** French, 5 students take **only** Spanish and 2 students take **both** French and Spanish.

Step 4 The choice of 2 students taking **both** classes will be correct only if the total number of students taking one or both classes is 13. The picture shows a total of 15 students (8 + 2 + 5 = 15), therefore the choice of 2 students taking both classes is **incorrect**.

Step 5 Choose another possible answer. Now the choice of 4 students taking both classes will be made. Write a 4 in the overlapping section.

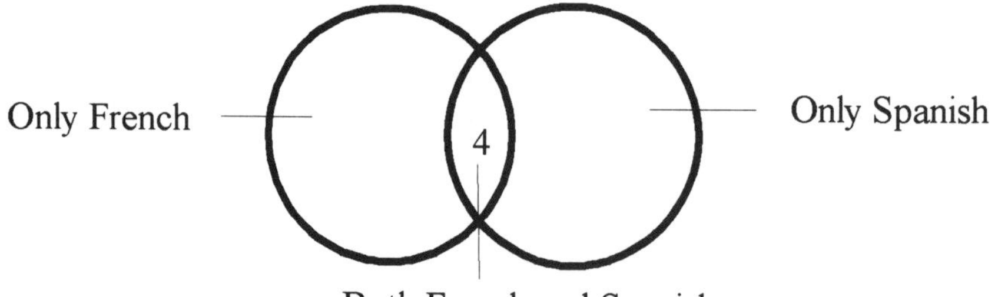

Step 6 Fill in the remaining number of students taking **only** one class or the other. Remember that the total number of students in the French class **must** be 10 and the total number of students in the Spanish class **must** be 7. If four students take both classes, then the number of students taking **only** French is 6 (10 – 4 = 6) and the number of students taking **only** Spanish is 3 (7 – 4 = 3).

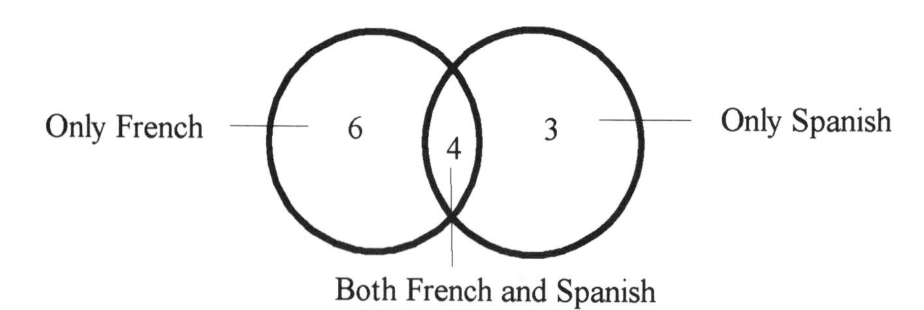

Step 7 The picture now shows that the French class has 10 students (6 + 4 = 10) and that the Spanish class has 7 students (3 + 4 = 7). The picture also shows that 6 students take only French, 3 students take **only** Spanish and 4 students take **both** French and Spanish. If the sum of these three numbers is 13, then the correct answer has been found. 6 + 4 + 3 = 13, therefore the choice of **4 students** taking **both** French and Spanish is correct.

Exercises:

1) In the art class, there are 16 students. In the math class, there are 7 students. If 5 students are in **both** classes, what is the total number of students in one or both classes?

2) In the French class, there are 52 students. In the Italian class, there are 37 students. If 10 students are in **both** classes, what is the total number of students in one or both classes?

3) In the art class, there are 30 students. In the history class, there are 27 students. If 6 students are in **both** classes, what is the total number of students in one or both classes?

4) Of 20 students, each takes French or Spanish or both. 15 students are in the French class and 9 students are in the Spanish class. How many students are in **both** classes?
Choices: 0, 2, 4, 5, 6

5) Of 12 students, each takes math or history or both. 10 students are in the math class and 3 students are in the history class. How many students are in **both** classes?
Choices: 0, 1, 2, 3, 4

6) Of 22 students, each takes history or math or both. 12 students are in the history class and 12 students are in the math class. How many students are in **both** classes?
Choices: 0, 1, 2, 5, 12

Shortcut #34 Intersections of four categories

An <u>intersection</u> of categories is the quantity in two or more categories at the same time. Draw four squares and label the rows and columns with the given categories and totals. Fill in the given numerical values and then determine other numerical values as needed to answer the question.

Example **In a group of 200 high school juniors and seniors, 70 are boys, 50 are senior boys and 80 are seniors. How many girls are juniors?**

Step 1 The four categories are: 1) junior boys 2) junior girls 3) senior boys and 4) senior girls. Set up the following chart and fill in the given information. Notice that totals **must** be shown for rows as well as columns.

	Juniors	Seniors	Total
Boys	**Junior Boys**	**Senior Boys**	**Boys**
Girls	**Junior Girls**	**Senior Girls**	**Girls**
Total	**Juniors**	**Seniors**	**Total Students**

➔

	Juniors	Seniors	Total
Boys		50	70
Girls			
Total		80	200

Step 2 Determine the number of junior boys: 70 – 50 = 20 junior boys.

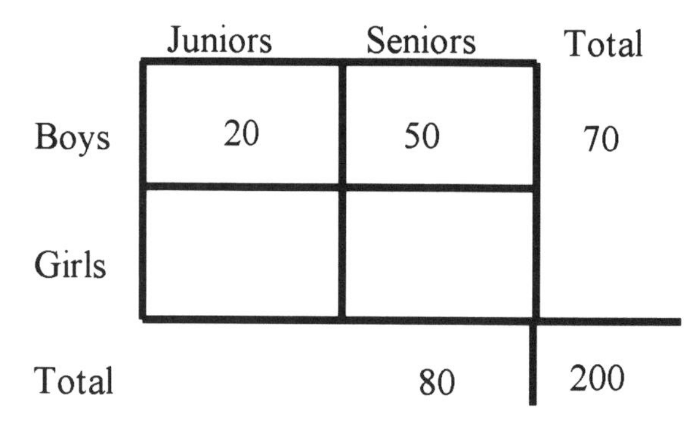

	Juniors	Seniors	Total
Boys	20	50	70
Girls			
Total		80	200

Step 3 Determine the total number of juniors: $200 - 80 = 120$ juniors.

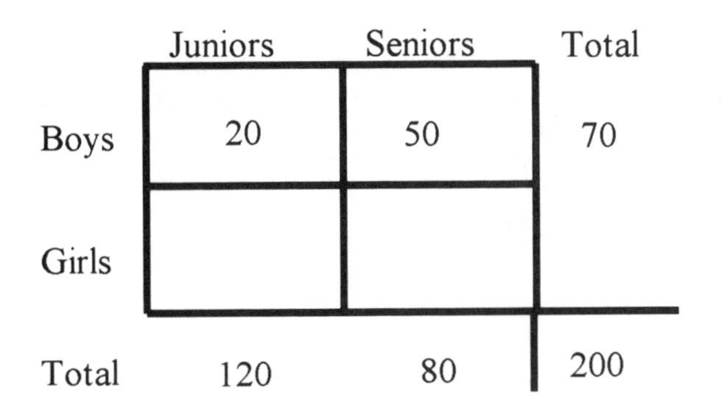

	Juniors	Seniors	Total
Boys	20	50	70
Girls			
Total	120	80	200

Step 4 Determine the number of junior girls: $120 - 20 = $ **100 junior girls** .

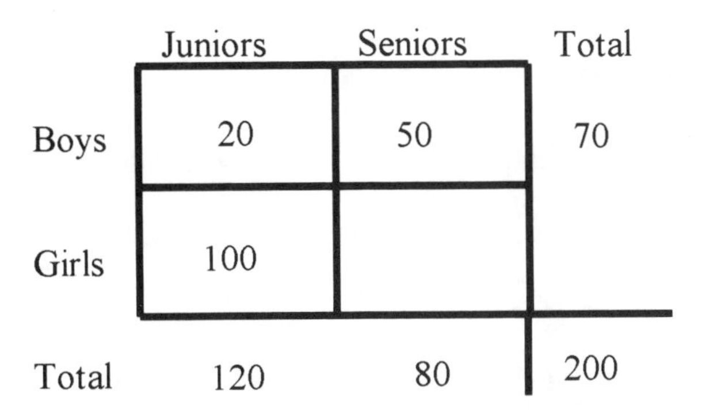

	Juniors	Seniors	Total
Boys	20	50	70
Girls	100		
Total	120	80	200

Note: This problem could alternatively been solved by determining the total number of girls, the number of senior girls and then subtracting these two numbers.

Exercises:

1) In a group of 170 freshmen and sophomores, 80 are boys, 35 are sophomore boys and 60 are sophomores. How many are freshmen girls?

2) In a group of 150 adults and children, 90 are children, 40 are female children and 70 are females. How many are adult males?

3) In a group of 250 high school and college students, 120 are high school students, 150 are girls and 90 are high school girls. How many college students are boys?

4) In a group of 200 high school and college students, 150 are high school students, 100 are girls and 80 are high school girls. How many college students are boys?

Appendix 1 Geometry Formulas

1) Area of a Rectangle or Square: **$A = L w$**

A = area **L = length** **w = width**

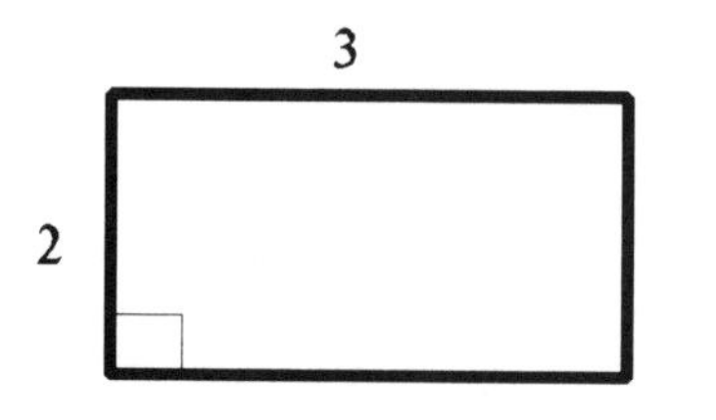

$A = L w$
$A = (3)\ (2)$
$A = 6$

Note: The length and width of a square are equal.

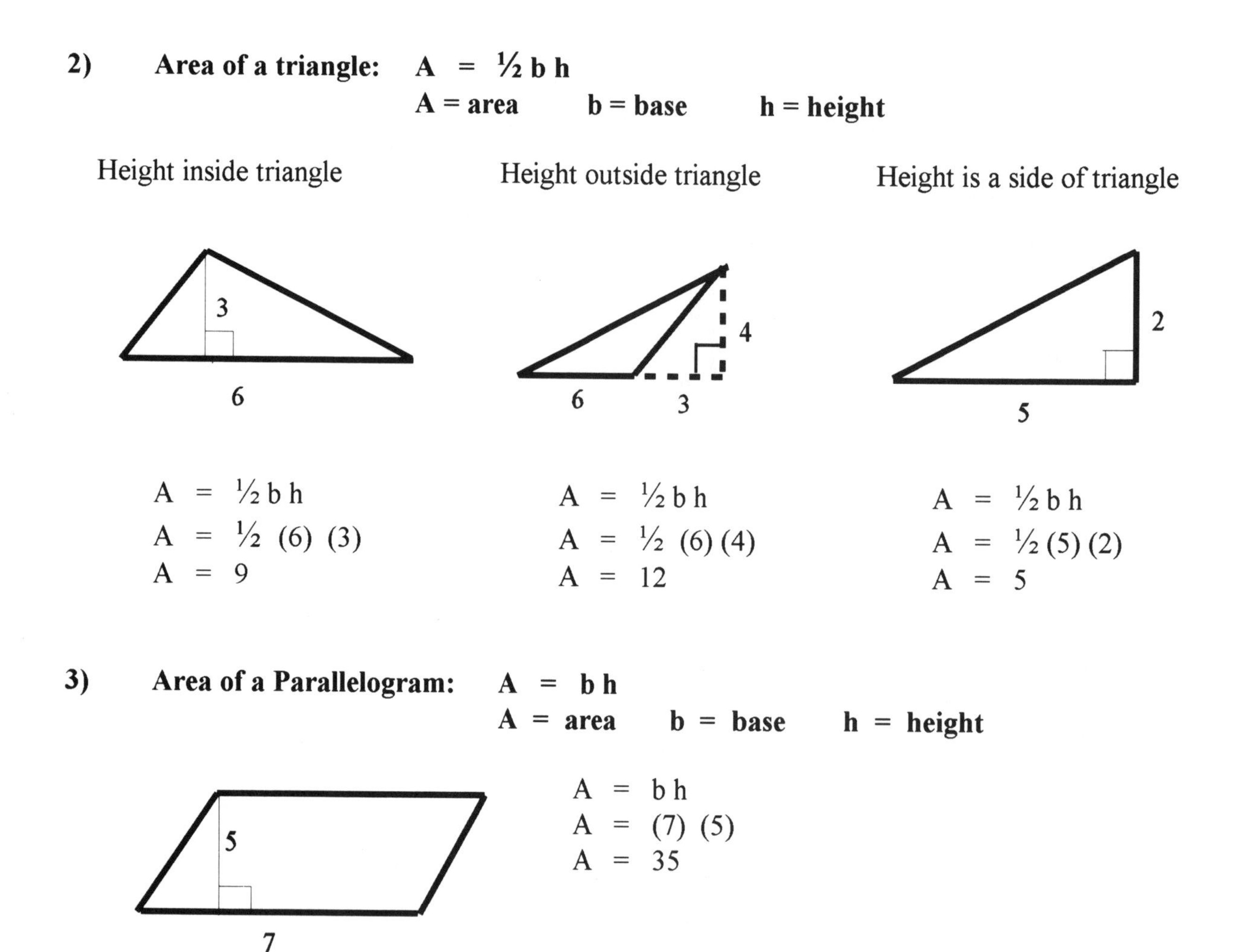

2) Area of a triangle: **$A = \frac{1}{2} b h$**

A = area **b = base** **h = height**

Height inside triangle

$A = \frac{1}{2} b h$
$A = \frac{1}{2}\ (6)\ (3)$
$A = 9$

Height outside triangle

$A = \frac{1}{2} b h$
$A = \frac{1}{2}\ (6)\ (4)$
$A = 12$

Height is a side of triangle

$A = \frac{1}{2} b h$
$A = \frac{1}{2}\ (5)\ (2)$
$A = 5$

3) Area of a Parallelogram: **$A = b h$**

A = area **b = base** **h = height**

$A = b h$
$A = (7)\ (5)$
$A = 35$

4) **Area of a Circle:** $A = \pi r^2$

A = **area** π = **pi (about 3.14 or 22/7)** r = **radius**

Note: If an exact value for the area of a circle is required do **not** substitute a numerical value for pi (π) because the exact value for π can **not** be represented numerically.

$A = \pi r^2$

$A = \pi (3)^2$

$A = \pi 9$ or 9π

5) **Circumference of a Circle:** $C = \pi d$

C = **circumference** π = **pi (about 3.14 or 22/7)**

d = **diameter**

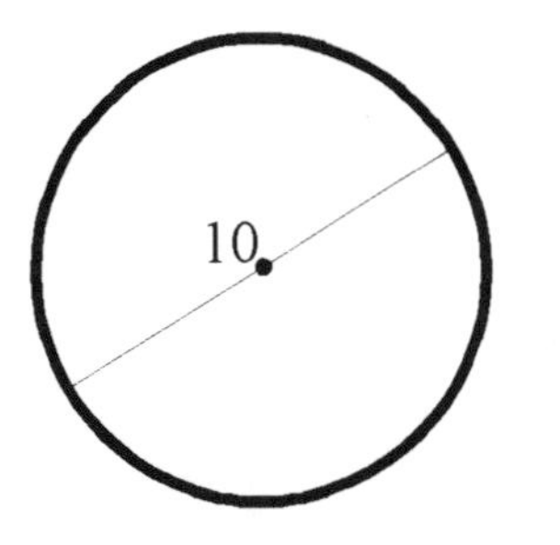

Note: If the exact value for the circumference is required do **not** substitute a numerical value for pi (π) because the exact value for pi (π) can **not** be represented numerically.

$C = \pi d$

$C = \pi (10)$ or 10π

6) **Area of a Sector of a Circle** $= \dfrac{\text{(central angle)}}{360°}$ x **(total area of the circle)**

Note: A **sector of a circle** is the region bounded by two radii and an arc of the circle.

Note: A **central angle** has its' vertex at the center of a circle. In the picture below, $\angle BAC$ is a central angle because its' vertex (point A) is the center of the circle.

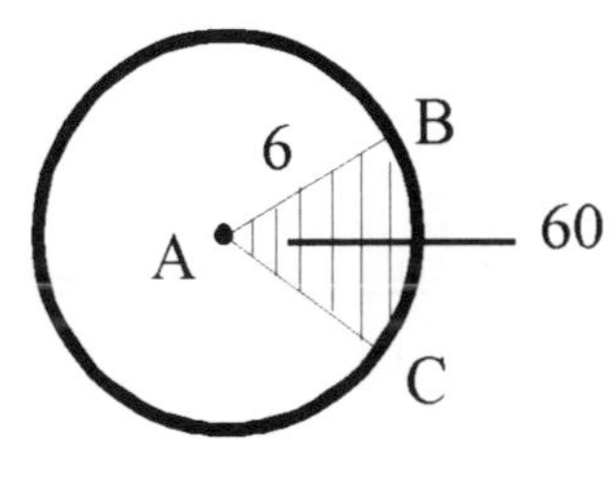

Center of circle at point A
Sector BAC is the shaded region
Arc BC and radii AB and AC are the boundaries of sector BAC
Central angle $\angle BAC = 60°$
Radius of Circle A is 6

Area of sector BAC (area of shaded region) $= \dfrac{(60°)}{(360°)} \times (36\pi) = 6\pi$

7) The Arc of a Circle can be measured in: 1) length or 2) degrees

a) Arc measured by length $= \dfrac{\text{(central angle)}}{360°}$ **x (circumference of the circle)**

From Circle A in #6: Length of arc BC $= \dfrac{(60°)}{(360°)} \times (12\pi) = 2\pi$

b) Arc measured by degrees = the degree measure of the central angle that intercepts the arc

From Circle A above in #6: Arc BC in degrees = 60°

Note: The entire arc (circumference) of a circle has a degree measure of 360°

8) Volume of a rectangular solid: V = L w h

V = volume L = length w = width h = height

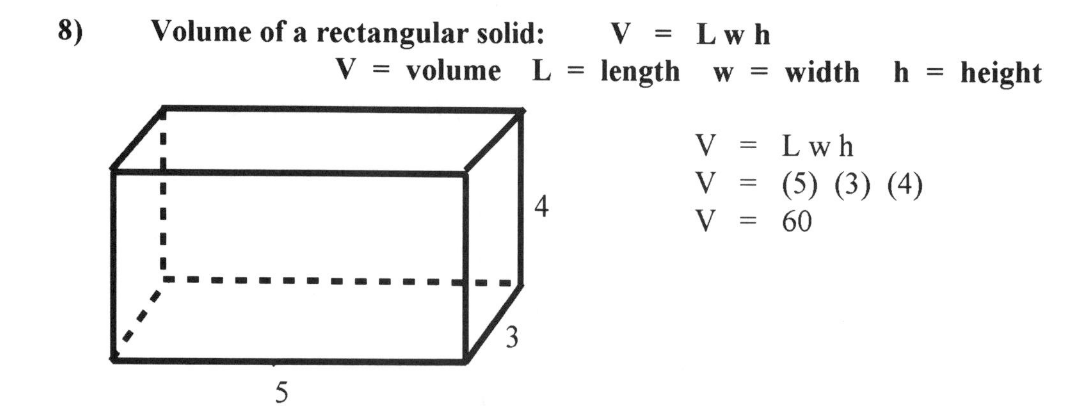

$V = L w h$
$V = (5)\ (3)\ (4)$
$V = 60$

9) <u>Surface Area</u> is the total area that surrounds a three dimensional object. For example, the surface area of a cube or rectangular solid is the sum of the areas of the: front, back, left, right, top and bottom faces (sides). The surface area of the rectangular solid above in #8 is:

Surface area = front + back + left + right + top + bottom

Surface area = 20 + 20 + 12 + 12 + 15 + 15 = 94

10) Perimeter of any Polygon: Sum of the length of all sides.

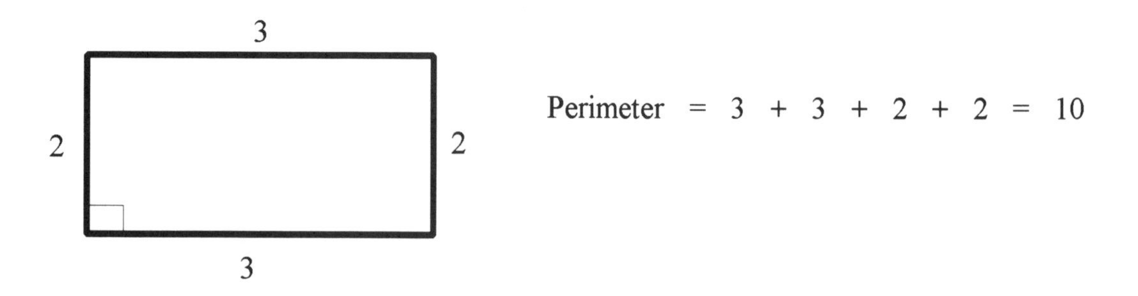

Perimeter = 3 + 3 + 2 + 2 = 10

11) Pythagorean Theorem: **The formula used to determine the length of any side of a right triangle. The formula applies if: 1) the length of any two sides is known or 2) if the triangle is an isosceles right triangle and the length of at least one side is known.**

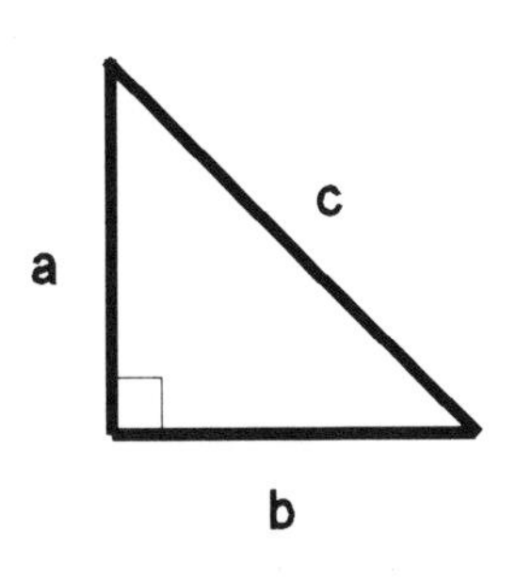

$a^2 + b^2 = c^2$

"a" and "b" are the legs, c = hypotenuse

Note: Either leg could be "a" or "b".

Note: When applying the formula to an isosceles right triangle, the same variable may represent each leg.

12) Special right triangles (45°, 45°, 90° and 30°, 60°, 90°):

In the these two right triangles, the length of just one side is enough information in order to determine the length of the other sides. The sides of these two triangles are always in the following proportions:

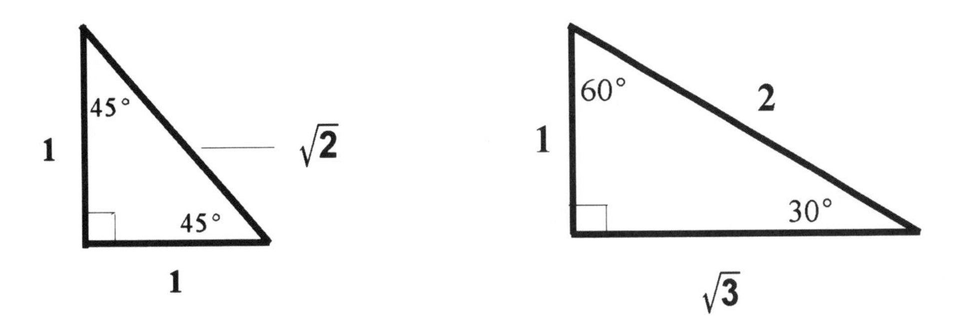

In any **isosceles right triangle** (45°,45°,90°), both legs are always the same length. The length of the hypotenuse is determined by multiplying either leg by $\sqrt{2}$. Conversely, the length of either leg is determined by dividing the hypotenuse by $\sqrt{2}$.

In any 30°, 60°, 90° right triangle, the length of the hypotenuse is always double the side opposite the 30° angle. The length of the side opposite the 60° angle is determined by multiplying the side opposite the 30° angle by $\sqrt{3}$. Conversely, the length of the side opposite the 30° angle is determined by dividing the side opposite the 60° angle by $\sqrt{3}$ or by taking half of the hypotenuse.

Appendix 2 Geometry Concepts

1) **If two straight lines intersect, then the opposite angles are equal. The angles opposite each other are known as vertical angles.**

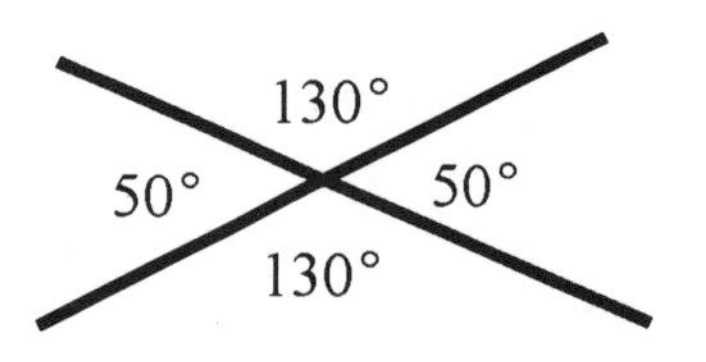

2) **The sum of all three angles of any triangle is 180° .**

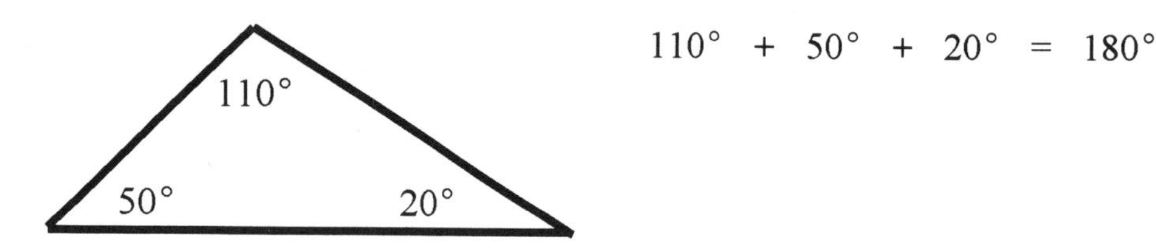

3) **The sum of all the angles around one side of a straight line is 180°.**

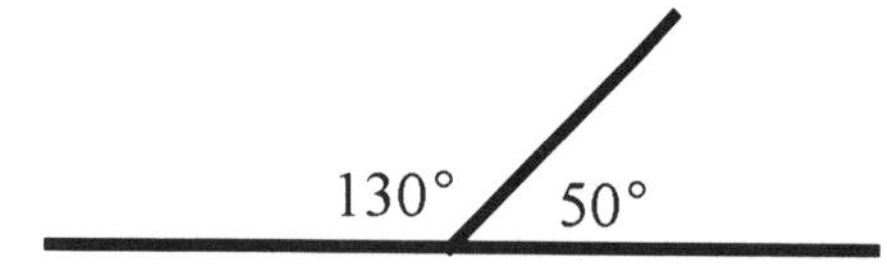

130° + 50° = 180°

4) **The sum of all the angles around a point or center of a circle is 360°.**

130° + 70° + 160° = 360°

120° + 70° + 170° = 360°

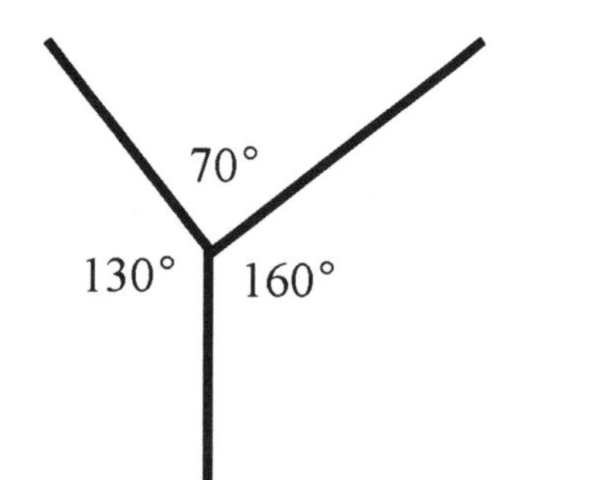

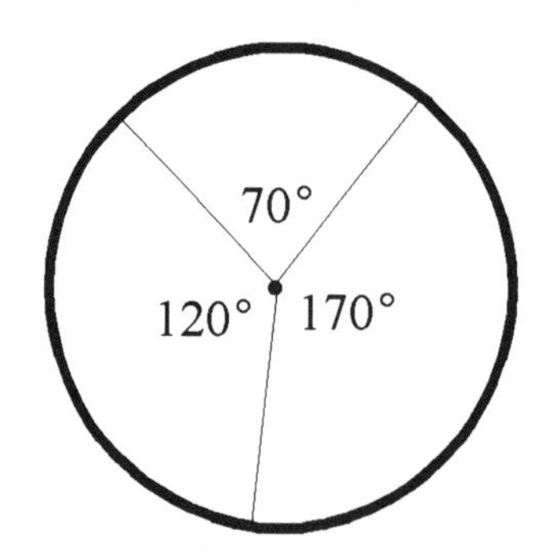

5) **The base angles of an isosceles triangle are the two angles opposite the two equal sides of an isosceles triangle. The base angles of an isosceles triangle are equal. Conversely, if two angles of a triangle are equal, then their opposite sides are equal and the triangle is an isosceles triangle.**

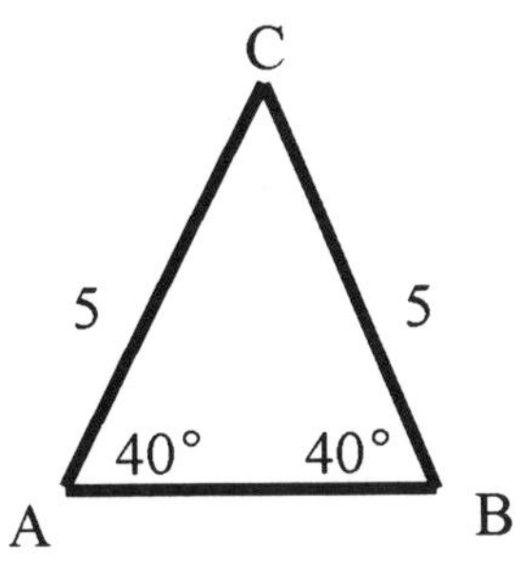

a) ∠A and ∠B are the base angles

b) ∠A is opposite side BC
∠B is opposite side AC

6) **In an isosceles triangle, the vertex angle is formed by the two equal sides and the base is the side opposite the vertex angle. The height (altitude) of any triangle is a line perpendicular to one side of the triangle and has one endpoint at the vertex opposite that side. The height of an isosceles triangle from the vertex angle, bisects the base and the vertex angle of the isosceles triangle.**

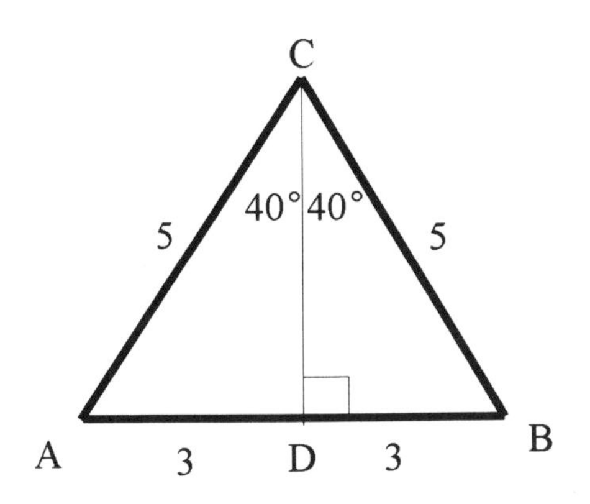

∠ACB is the vertex angle
AB is the base
CD is the height from the vertex angle
Point C is the vertex of ∠ACB
AD = BD (base bisected)
∠ACD = ∠BCD (vertex angle bisected)

7) **A quadrilateral is a closed four sided figure.**
The sum of all four angles of any quadrilateral is 360°.

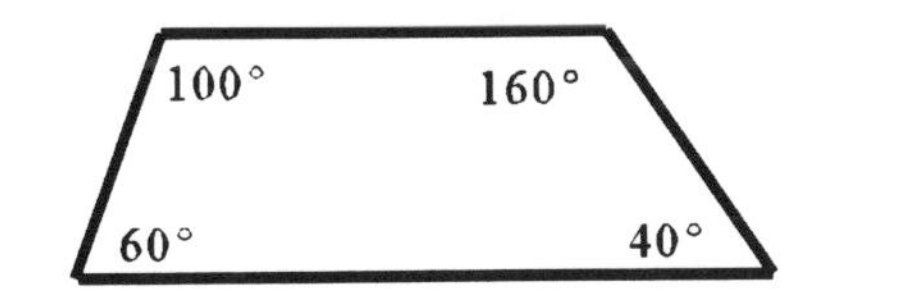

160° + 100° + 60° + 40° = 360°

8) **The diagonals of a rectangle are equal and bisect each other.**

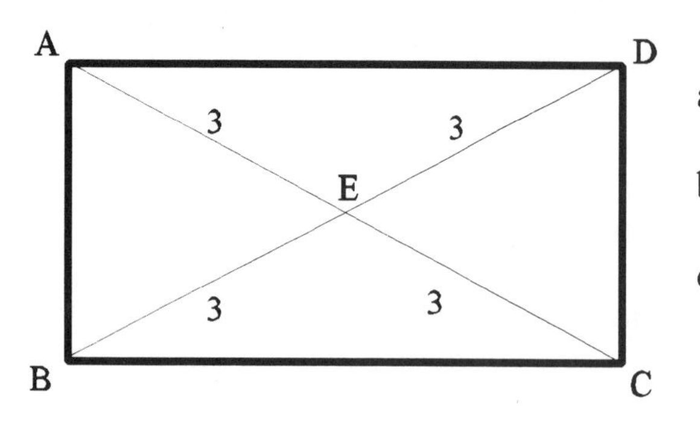

a) AC and BD are the diagonals

b) AC = BD (diagonals are equal)

c) BE = ED , AE = EC
(diagonals bisect each other)

9) **Properties of a Parallelogram:**

a) **Opposite sides are equal and parallel.**

AB = CD, AD = BC, AB ∥ CD, AD ∥ BC

b) **Opposite angles are equal.**

$\angle A = \angle C$, $\angle B = \angle D$

c) **Any pair of consecutive angles of a parallelogram are supplementary (sum of two angles equals 180°). Consecutive angles of a polygon are two angles that are next to each other.**

$\angle A + \angle B = 180°$, $\angle B + \angle C = 180°$
$\angle C + \angle D = 180°$, $\angle A + \angle D = 180°$

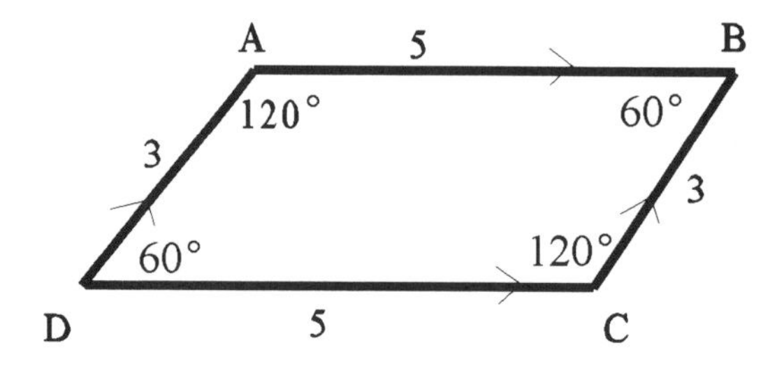

10) **If two parallel lines are intersected by another line, then the following angle relationships exist.**

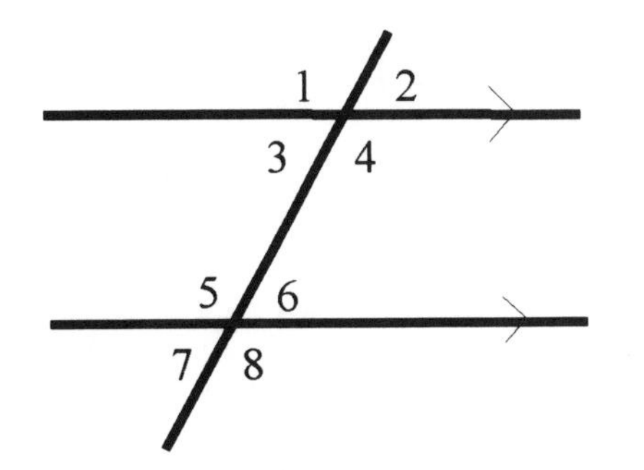

Corresponding angles:

$\angle 1 = \angle 5$
$\angle 3 = \angle 7$
$\angle 2 = \angle 6$
$\angle 4 = \angle 8$

Vertical angles:

$\angle 1 = \angle 4$
$\angle 2 = \angle 3$
$\angle 5 = \angle 8$
$\angle 6 = \angle 7$

Alternate interior angles:

$\angle 3 = \angle 6$
$\angle 4 = \angle 5$

Alternate exterior angles:

$\angle 1 = \angle 8$
$\angle 2 = \angle 7$

Same side interior angles are supplementary:

$\angle 4 + \angle 6 = 180°$
$\angle 3 + \angle 5 = 180°$

Example:

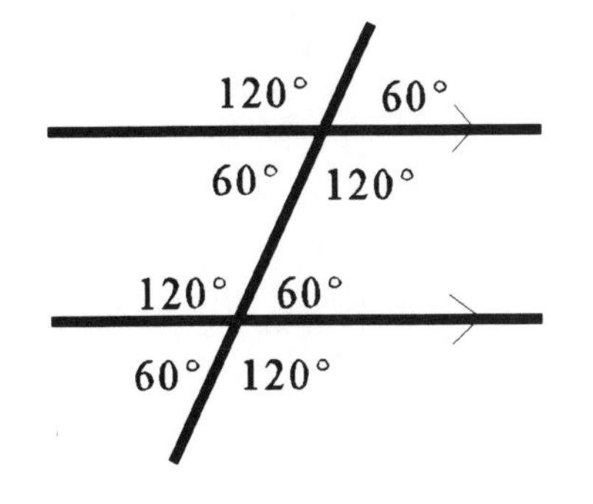

Appendix 3 Miscellaneous Formulas and Topics

1) **Probability** $= \dfrac{\textbf{number of desired items}}{\textbf{total number of items}}$

Note: A probability may be in the form of a fraction or percent.

Example A jar contains 3 red marbles, 2 blue marbles and 5 white marbles. What is the probability of choosing a red marble on the first try without looking?

$$\text{probability} = \frac{\text{number of red marbles}}{\text{total number of marbles}} = \frac{\mathbf{3}}{\mathbf{10}} \quad \textbf{or 30\%}$$

2) **Distance of a moving object:** $\mathbf{D = R\ T}$

D = distance R = average rate (or average speed) T = time

Note: Rate (or speed) and time must use the same unit of time. Usually this is hours. An example of rate or speed might be 50 mph (mile per hour) which indicates that the object will travel 50 miles in 1 hour if the speed is constant at 50 mph.

Note: Average rate (or average speed) must be used because an object may travel at different speeds during a trip.

Example What is the distance a car travels if the average speed is 50 mph for 3 hours?

$$D = R\,T$$

$$D = (50 \text{ mph}) \times (3 \text{ hours})$$

$$D = 150 \text{ miles}$$

3) **Average Speed** $= \dfrac{\textbf{Total distance}}{\textbf{Total time}}$

Example A car travels 10 miles in 1 hour and then travels 50 miles in 2 hours. What is the average speed?

$$\text{average speed} = \frac{10 \text{ miles} + 50 \text{ miles}}{1 \text{ hour} + 2 \text{ hours}}$$

$$\text{average speed} = \frac{60 \text{ miles}}{3 \text{ hours}} = \textbf{20 mph}$$

4) **A <u>weighted average</u> is the average of two or more groups in which the number of items in the groups is not all the same.**

Example Of a class of 25 students, 10 students received a grade of 80 and the other 15 students received a grade of 90. What was the average grade of the class?

A weighted average must be used because a different number of students received grades of 80 and 90. The weights will be the number of students. The grade of 80 has a weight of 10 because 10 students received a grade of 80. The grade of 90 has a weight of 15 because 15 students received a grade of 90. Multiply each grade by its respective weight and then divide this sum by the sum of the weights (25 students).

$$\text{average grade} = \frac{(80)\ (10\ \text{students}) + (90)\ (15\ \text{students})}{25\ \text{students}}$$

$$\text{average grade} = \frac{800 + 1350}{25} = \mathbf{86}$$

Note: The average grade is **not** 85 because the number of students receiving an 80 and 90 is **not** equal. If the same number of students received an 80 and 90, then the average grade would have been 85.

5) **The three forms of a division problem and answer:**

A division problem may be written in any of the following three forms. The following three examples represent the same division problem, stated as: 7 divided by 5 or 5 divided into 7.

$$7 \div 5 = \qquad \text{or} \qquad 5\overline{)7} \qquad \text{or} \qquad \frac{7}{5} =$$

The answer to the above division problem may be in the following three forms: remainder, fractional or decimal. "r" represents the word "remainder."

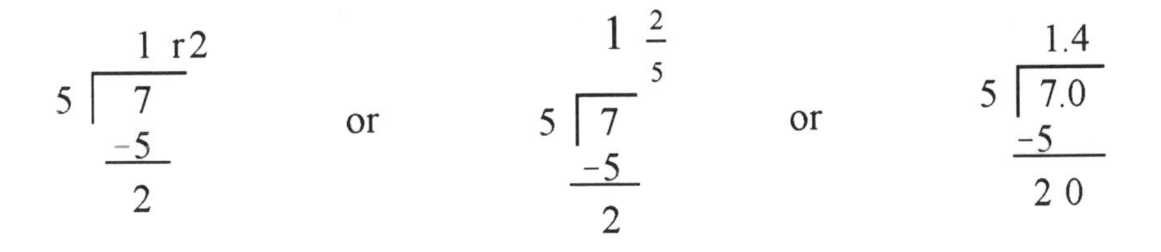

Glossary

adjacent: next to each other

adjacent angles: two angles that share a common vertex and a common side

altitude: (see height)

arc: part of the circumference of a circle. An arc can be measured in length or degrees. An arc's measure in degrees is equal to the central angle that intercepts that arc. The entire arc of a circle (circumference) measures 360°.

area: the surface covered by a two dimensional shape such as a rectangle or circle

arithmetic mean: (see average)

arrangement: a group of items. A new arrangement is created if one or more items in the arrangement are changed **or** if the order of the same items in the arrangement is changed. The total number of arrangements is determined by **multiplying** a series of numbers together.

average: (also known as mean or arithmetic mean) the result obtained by dividing the **sum** of two or more quantities by the number of quantities.

base angles of an isosceles triangle: the two equal angles. Each base angle is opposite one of the equal sides of the isosceles triangle.

base of an isosceles triangle: the unequal side of an isosceles triangle. The base is the side opposite the vertex angle. The two equal sides of an isosceles triangle are known as the legs.

base of a power: In the example, 2^3, 2 is the base and 3 is the exponent. $2^3 = 2 \times 2 \times 2 = 8$

base of a triangle: is a side of a triangle (or any polygon) that the height intersects. In the formula for the area of a triangle, $\mathbf{A = \frac{1}{2} b h}$, b = base and h = height.

bases of a trapezoid: the two parallel sides of a trapezoid. The two non-parallel sides of a trapezoid are the legs.

bisect: to divide or cut in half

central angle: an angle whose vertex is the center of a circle

chord: a line segment whose endpoints are on a circle

circumference: the total distance around a circle. The formula for circumference is: $C = \pi d$, where, C = circumference, π (pi) is approximately 3.14 or 22/7 and d = diameter.

circumscribed: a circle is circumscribed around a polygon if each vertex of the polygon is on the circle. A polygon is circumscribed around a circle if each side of the polygon is tangent to the circle.

coefficient: a number that is multiplied by one or more variables (letters). In the example, 6xy , 6 is the coefficient.

combination: a group of items. A new combination is created **only** if one or more items in the combination are changed. A different order of the **same** items does **not** create a new combination. The total number of combinations that contain **only** two items is determined by **adding** a series of numbers.

common denominator: the identical denominator that fractions may have. A common denominator always exists for any group of fractions.

common factor: an identical factor of two or more numbers or terms

complementary angles: any two angles whose sum is $90°$

complex fraction: a fraction in which either the numerator, denominator or both contain fractions

concentric circles: circles that have the same center

consecutive angles: any two angles of a polygon that are next to each other

consecutive numbers: numbers that follow in order without interruption. Consecutive numbers may or may not continue indefinitely.
Examples: Consecutive integers: 1, 2, 3, 4
Consecutive even integers: 2, 4, 6
Consecutive odd integers: 5, 7, 9, 11

constant: a term that is a number and does **not** contain any variables, for example in the expression $3x + 8$, 8 is the constant.

cross-multiply: the first step in solving a proportion. The product of the numerator of one fraction and the denominator of the other fraction is set equal to the product of the other numerator and denominator.

cube: a three dimensional shape in which all six faces (sides) are squares and form right angles at the corners

cube root: is denoted by a radical sign with a 3 at the upper left ($\sqrt[3]{\;\;}$). The cube root of a number is a number that when multiplied by itself three times equals the number inside the cube root sign. $\sqrt[3]{8} = 2$ because $2 \times 2 \times 2 = 8$. $\sqrt[3]{-8} = -2$ because $(-2)(-2)(-2) = -8$.

degree: unit of measure of an angle or arc of a circle.

denominator: the quantity below the line in a fraction

diagonal: a line segment whose endpoints are non-consecutive vertices of a polygon

diameter: a line segment that passes through the center of a circle and whose endpoints are on the circle

difference: indicates subtraction

digit: a single number. The ten digits are: 0, 1, 2, 3, 4, 5, 6, 7, 8, and 9

distinct: describes different numbers or items

dividend: a quantity that is divided by another quantity. In the example, $10 \div 5 = 2$ (also written as $5\overline{)10}$ with quotient 2, or $\frac{10}{5} = 2$) 10 is the dividend, 5 is the divisor and 2 is the quotient (answer).

divisible: division in which there is **no** remainder

divisor: a quantity that is divided **into** another quantity. See dividend for an example.

edge: a line segment formed by the intersection of two faces (sides) of a three dimensional object

endpoint: the point at the end of a line segment

equation: a statement in which two quantities are set equal to each other. Generally, an equation contains one variable (letter) whose numerical value can be determined.

equilateral triangle: a triangle in which all three sides are equal and all three angles equal $60°$.

exponent: in the example 2^3, 3 is the exponent and 2 is the base. $2^3 = 2 \times 2 \times 2 = 8$

expression: the word used to describe a term or group of terms that are added or subtracted, for example, $3x - 7$ is an expression that contains two terms.

face: a side of a three dimensional shape. For example, the 6 faces of a rectangular solid are: top, bottom, front, back, left, and right.

factor: 1) a number that is divisible into another number, for example 5 is a factor of 10 because 10 is divisible by 5. 2) the inverse process of multiplication, for example, 10 can be factored as 5 x 2.

height: (also known as altitude) a line segment in which one endpoint is the vertex of a polygon and the line segment is perpendicular to the side opposite that vertex. In the area formula of a triangle, $\mathbf{A = \frac{1}{2} b h}$, h = the height and b = the base.

hypotenuse: the side of a right triangle that is opposite the right angle. The hypotenuse is always the longest side of a right triangle. The other two sides of a right triangle form the right angle and are called the legs.

improper fraction: a fraction in which the numerator is greater than the denominator

inclusive: indicates that the first and last numbers of a group are to be included. For example, the numbers 2 through 5 inclusive are: 2, 3, 4, 5.

inequality: a comparison of two quantities:
$x > 5$ (or $5 < x$), x is greater than 5
$x < 5$ (or $5 > x$), x is less than 5
$x \geq 5$ (or $5 \leq x$), x is greater than or equal to 5
$x \leq 5$ (or $5 \geq x$), x is less than or equal to 5

inscribed: a circle is inscribed in a polygon if each side of the polygon is tangent to the circle. A polygon is inscribed in a circle if each vertex of the polygon is on the circle.

integer: a positive or negative whole number or zero. Zero is neither positive nor negative. Examples of integers are: ... −3, −2, −1, 0, 1, 2, 3...

irrational numbers: numbers that are **not** rational. Irrational numbers can **not** be expressed as a fraction in which the numerator and denominator are integers and the denominator does **not** equal zero. Examples of irrational numbers are: π, and square roots that can **not** be simplified into a whole number such as $\sqrt{2}$ or $\sqrt{3}$.

isosceles right triangle: a triangle that contains a right angle (90°) and whose legs are equal in length

isosceles trapezoid: a trapezoid in which the two non-parallel sides (legs) are equal

isosceles triangle: a triangle in which two sides are equal

least common denominator (LCD): the lowest identical denominator that two or more fractions may have. A least common denominator always exists for any group of fractions.

legs of an isosceles triangle: the two equal sides of an isosceles triangle. The third (unequal side) is the base.

legs of a right triangle: the two shorter sides of a right triangle that form the right angle. In the Pythagorean Theorem, $a^2 + b^2 = c^2$, a and b represent the legs of the right triangle.

legs of a trapezoid: the two non-parallel sides

less than: indicates a subtraction, for example, 5 less than x means $x - 5$ and x less than 5 means $5 - x$

line: a series of straight, connected points that extend in two directions without ending

line segment: a series of straight, connected points that has two endpoints

mean: (see average)

median: the middle number in a series of numbers that are written in order, for example, in the series of numbers: 1, 7, 15, 19, 23; 15 is the median

midpoint: a point that bisects a line segment

minor and major arcs: a minor arc is labeled by the two endpoints of the arc. A minor arc is less than 180°. A major arc is labeled with three letters, the first and last being the endpoints of the arc and the middle letter being somewhere on the arc. A major arc is greater then 180°.

mixed number: a whole number combined with a fraction, for example 3½

mode: the number or value that appears most often in a group. A group of numbers may have more than one mode. Examples: In the group: 2, 3, 5, 5, 6; 5 is the mode.
In the group: 2, 2, 2, 3, 3, 4, 4, 4; 2 and 4 are the modes.

multiple: the multiples of a number are determined by multiplying that number by whole numbers. The multiples of 3 are: 0, 3, 6, 9, 12...

3 x 0	3 x 1	3 x 2	3 x 3	3 x 4
0	**3**	**6**	**9**	**12**

natural numbers: the positive integers: 1, 2, 3, 4...

numerator: the quantity above the line in a fraction

order of operations: the correct order of performing math operations. The order of operations must be performed from left to right and within parentheses first.
The order of operations is:
1st) simplify exponents
2nd) perform multiplication and division
3rd) perform addition and subtraction

origin: the intersecting point of the x-axis and y-axis of a coordinate plane. The (x,y) coordinates of the origin are (0,0).

parallel lines: lines that do **not** intersect. The symbol for parallel lines is $\|$.

parallelogram: a quadrilateral whose opposite sides are equal and parallel

perfect square: a number whose square root is a whole number, for example, 9 is a perfect square because $\sqrt{9} = 3$.

perimeter: the sum of the lengths of all sides of a polygon

perpendicular lines: lines that form a right angle (90°). The symbol for perpendicular lines is $\perp$.

pi (π): a Greek letter whose numerical value is approximately 3.14 or 22/7 .

polygon: a closed two dimensional figure formed by straight lines such as a square, rectangle or triangle

power: the value of a quantity that has an exponent. The value of the power 2^3 is 8 ($2^3 = 2 \times 2 \times 2 = 8$).

prime factor: a factor that is a prime number

prime number: an integer greater than 1 that is divisible by itself or 1 **only**. Some examples are: 2, 3, 5, 7, 11...

product: the answer to a multiplication problem. For example the product of 3 and 5 is 15.

proportion: an equation in which two fractions are set equal to each other.

Pythagorean Theorem: a formula used to determine the length of a side of a right triangle. The formula is $\mathbf{a^2 + b^2 = c^2}$, where a and b are the legs of the right triangle and c is the hypotenuse of the right triangle. The formula applies if the length of any two sides of the right triangle are known. The formula applies to an isosceles right triangle if the length of at least one side is known. When applying the formula to an isosceles right triangle, the same variable may represent each leg.

Pythagorean Triple: the possible length of sides of a right triangle. When applicable, Pythagorean Triples are a shortcut to the Pythagorean Theorem. Common Pythagorean Triples are: 3, 4, 5 and 5, 12, 13.

quadratic equation: An equation in which a variable (letter) contains an exponent of 2.

quadrilateral: a four sided polygon

quotient: the answer to a division problem. For example, the quotient of $15 \div 3$ is 5.

radii: plural for radius

radius: a line segment in which one endpoint is the center of a circle and the other endpoint is on the circle

rate: speed

ratio: the comparison of two quantities. A ratio can be represented in the following three ways: 2/3 , 2 : 3 , or 2 to 3 . Notice that any fraction is a ratio.

rational numbers: numbers that can be written as a fraction in which the numerator and denominator are integers and the denominator is **not** zero.

real numbers: the rational and irrational numbers

reciprocal: the inverted form of a fraction. The reciprocal of 2/3 is 3/2, the reciprocal of 2 is 1/2.

rectangular solid: a three dimensional shape in which all six faces (sides) are rectangles and form right angles at each corner

regular polygon: a polygon in which all sides are equal and all angles are equal

remainder: the left over quantity in the answer to a division problem. The remainder is preceded by "r". The remainder in the division problem, $7 \div 2$ is 1: $2\overline{)7}$ = 3 r 1

right angle: a 90° angle, indicated by the symbol ∟. A right angle is formed by two perpendicular lines.

right triangle: a triangle that contains a right angle (90°)

sector of a circle: the region of a circle that is completely bounded by two radii and an arc of the circle

semicircle: 1) half the area of a circle bounded by the diameter 2) an arc whose endpoints are on the diameter. This arc measures 180°.

square root: the square root of a number is a number that when multiplied by itself equals the number inside the square root symbol (√). For example, $\sqrt{9} = 3$, because $3 \times 3 = 9$. The square root symbol (√) represents the positive square root only. The complete square root of 9 is actually (+3, −3) because $3 \times 3 = 9$ and $-3 \times -3 = 9$. Square roots are simplified by factoring out perfect squares, thereby leaving the smallest number possible inside the square root. For example the simplified form of $\sqrt{12} = \sqrt{4} \times \sqrt{3} = 2\sqrt{3}$.

squaring: the process of applying an exponent of 2 to a quantity. For example, x squared equals x^2.

sum: the answer to an addition problem, for example the sum of 3 and 5 is 8

supplementary angles: any two angles whose sum is 180°

surface area: the total area surrounding a three dimensional shape. For example, the surface area of a cube or rectangular solid is the sum of the areas of the: front, back, top, bottom, left and right faces (sides).

tangent: a line is tangent to a circle or arc if the line intersects the circle or arc at **only** one point

term: the product of a group of numbers and variables (letters). Terms are separated by addition or subtraction signs. All terms may be multiplied or divided. Like terms are those that have identical variables (letters) and identical exponents to those variables. Like terms may be multiplied, divided, added or subtracted. The following two examples are the sum of two like terms: $5x + 6x = 11x$ and $3x^2y + 4x^2y = 7x^2y$. Terms that are **not** like terms are said to be unlike terms and may only be multiplied or divided.

trapezoid: a quadrilateral (four sided figure) in which **only** one pair of opposite sides are parallel. The two parallel sides of the trapezoid are the bases and the two non-parallel sides are the legs.

units digit: the digit in the ones column, for example in the number 532; 2 is the units digit

variable: a letter that may represent one or more numbers

vertex: the common endpoint of an angle

vertex angle of an isosceles triangle: the angle that is formed by the two equal sides. The other two angles are base angles. The vertex angle is opposite the unequal side (base).

vertical angles: angles formed by two intersecting lines. Vertical angles are opposite each other and are equal.

volume: the total space contained in a three dimensional object

whole numbers: the integers greater or equal to zero: 0, 1, 2, 3, 4...

Index

Answers to Exercises

Shortcut #1, page 1

1) yes **2)** yes **3)** yes **4)** no **5)** yes **6)** no

Shortcut #2, page 2

1) 55 (d) **2)** 111 (b) **3)** 888 (c) **4)** 22 (c) **5)** 333 (b) **6)** 333 (a)

Shortcut #3, page 3

1) 3/4 **2)** 7/10 **3)** 3/5 **4)** 4/9

Shortcut #4, page 4

1) 7/8 **2)** 12/13 **3)** 11/12

Shortcut #5, page 4

1) 5/9 **2)** 2/3 **3)** 5/8

Shortcut #6, page 5

1) 8 **2)** 12 **3)** 7 ½ or 7.5 **4)** 40 **5)** 45 **6)** 21 **7)** 7 ½ or 7.5 **8)** 50 **9)** 60

10) 17 ½ or 17.5 **11)** 18

Shortcut #7, page 6

1) 25.5 **2)** 50.5 **3)** 500.5 **4)** 1,000.5 **5)** 38.5 **6)** 250.5

Shortcut #8, page 7

1) 1,275 **2)** 5,050 **3)** 500,500 **4)** 2,001,000 **5)** 2,926 **6)** 125,250

Shortcut #9, page 8

1) 1,440 **2)** 2,200 **3)** 3,500

Shortcut #10, page 11

1) b **2)** b **3)** a **4)** a **5)** b **6)** b **7)** a **8)** a **9)** b **10)** c

Shortcut #11, page 13 - 14

1) b **2)** a **3)** b **4)** b **5)** b **6)** a **7)** b **8)** a **9)** b **10)** b **11)** a

Shortcut #12, page 24 - 26

1) d **2)** d **3)** d **4)** d **5)** d **6)** d **7)** d **8)** d **9)** a **10)** c **11)** d **12)** d

13) d **14)** d **15)** d **16)** d **17)** d **18)** d **19)** d **20)** d **21)** c **22)** c **23)** a

24) d **25)** c **26)** d **27)** d **28)** d **29)** d **30)** d

Shortcut #13, page 28

1) b **2)** c **3)** a **4)** a **5)** b **6)** a **7)** a **8)** a **9)** b **10)** c

Shortcut #14, page 31

1) a **2)** a **3)** b **4)** b **5)** a **6)** a **7)** a **8)** a **9)** a **10)** b

Shortcut #15, page 32 - 33

1) 15 **2)** -12 **3)** 30 **4)** 12 **5)** 21 **6)** 4 **7)** 1 **8)** 4 **9)** 3

Shortcut #16, page 35

1) 8 **2)** 3 **3)** 13 **4)** 14 **5)** -3 **6)** 3 **7)** 1 **8)** 7 **9)** -1

Shortcut #17, page 36

1) 3/10 **2)** 27/28 **3)** 8/15 **4)** 9/16

Shortcut #18, page 40

1) 65π **2)** 55π **3)** 18 **4)** $64 - 16\pi$ **5)** $100 - 25\pi$ **6)** $64 - 4\pi$

7) $36 - \frac{9\pi}{2}$ **8)** $\pi - 2$ **9)** $\frac{25\pi - 50}{4}$

Shortcut #19, page 43 - 44

1) y **2)** z **3)** x **4)** y **5)** x **6)** y **7)** $\angle B > \angle C > \angle A$ **8)** $\angle A > \angle B > \angle C$

9) 50 **10)** 45 **11)** 70 **12)** 4 **13)** 5

Shortcut #20, page 45

1) $3 < x < 7$ **2)** $4 < x < 8$ **3)** $9 < x < 23$ **4)** $8 < x < 18$ **5)** $1 < x < 15$

6) $2 < x < 4$ **7)** $2 < x < 6$ **8)** $0 < x < 20$

Shortcut #21, page 46

1) yes **2)** no **3)** yes **4)** no **5)** yes **6)** yes **7)** yes **8)** no

Shortcut #22, page 50 - 51

1) 70 **2)** 120 **3)** 5 **4)** 6 **5)** 4 **6)** 3 **7)** 1 **8)** 48 **9)** 160 **10)** 64 **11)** 100

Shortcut #23, page 54 - 55

1) 5 **2)** 4 **3)** 13 **4)** 5 **5)** 8 **6)** 12 **7)** 6 **8)** 50

9) $\sqrt{34}$ (Pythagorean Triple is **not** applicable) **10)** $2\sqrt{34}$ (Pythagorean Triple is **not** applicable)

Shortcut #24, page 57 - 58

1) $110 < y < 130$ **2)** $100 < y < 150$ **3)** $55 < y < 70$ **4)** $60 < y < 80$

Shortcut #25, page 59 - 60

1) 5 **2)** 31 **3)** 5 **4)** 20 **5)** 30 **6)** 30 **7)** 32

Shortcut #26, page 61 - 62

1) 12 **2)** 18 **3)** 35/6 **4)** 1/6 **5)** 7/18 **6)** 9/8 **7)** 1/20 **8)** 3 **9)** 15 **10)** 3.5

11) 1.035 **12)** .05 or 5% **13)** .21 or 21% **14)** .5

Shortcut #27, page 64

1) $\frac{7}{3}$ **2)** 4 **3)** $\frac{1}{3}$ **4)** 85% **5)** 125% **6)** 25%

Shortcut #28, page 65

1) 5 **2)** 10 **3)** 57 **4)** 90 **5)** 965 **6)** 373

Shortcut #29, page 67

1) 6 **2)** 120 **3)** 720 **4)** 6 **5)** 24 **6)** 12 **7)** 24 **8)** 30 **9)** 360

Shortcut #30, page 69

1) 6 **2)** 28 **3)** 10 **4)** 3 **5)** 15

Shortcut#31, page 70

1) 72 **2)** 60

Shortcut #32, page 73 - 74

1) 420 **2)** 2 ¼ or 2.25 **3)** 10,800 **4)** 1/10 or .1 **5)** 180/d **6)** $\frac{1{,}000d}{c}$ **7)** $\frac{500d}{c}$

Shortcut #33, page 77

1) 18 **2)** 79 **3)** 51 **4)** 4 **5)** 1 **6)** 2

Shortcut #34, page 79

1) 65 **2)** 30 **3)** 70 **4)** 30